122 46 3396

The Intelligent Network
Standards

Other McGraw-Hill Communications Books of Interest

The Intelligent Network Standards

Their Application to Services

Igor Faynberg

Lawrence R. Gabuzda

Marc P. Kaplan

Nitin J. Shah

AT&T Bell Laboratories

McGraw-Hill

New York   San Francisco   Washington, D.C.   Auckland   Bogotá
Caracas   Lisbon   London   Madrid   Mexico City   Milan
Montreal   New Delhi   San Juan   Singapore
Sydney   Tokyo   Toronto

Library of Congress Cataloging-in-Publication Data

The Intelligent network standards: Their application to service / Igor
  Faynberg... [et al.].
      p.   cm.—(McGraw-Hill series on telecommunications)
    ISBN 0-07-021422-0
    1. Wide area networks (Computer networks)   2. Integrated services
  digital networks.   I. Faynberg, Igor.   I. Series.
  TK5105.7.I566   1997
  004.6′7—dc20   96-43806
                                           CIP

McGraw-Hill

A Division of The **McGraw·Hill** Companies

1 2 3 4 5 6 7 8 9 0   BKP/BKP   9 0 1 0 9 8 7 6

ISBN 0-07-021422-0

*The sponsoring editor for this book was Stephen S. Chapman, the edit-
ing supervisor was David E. Fogarty, and the production supervisor
was Suzanne W. B. Rapcavage. It was set in Century Schoolbook by
Donald A. Feldman of McGraw-Hill's Professional Book Group composition unit.*

Printed and bound by Quebecor / Book Press.

This book was printed on acid-free paper.

To
Anne Bishop Faynberg
Aida and Nicholas Gabuzda
Lynne, Zak, Elissa, and Eddie Kaplan
Nita, Shrenik, and Shreepal Shah

Contents

Foreword

To casual readers, the term "Intelligent Network" may well conjure up visions of some sort of electronic IQ spread liberally across the landscape. But to networking professionals, the IN concept means something else entirely. In practical terms, it connotes an ability to provide the broadest possible range of sophisticated services and features in a distributed environment. A tall order. In order to allow designers to realize this ambitious goal, IN architectures must provide a layered environment, one which separates services and features—as well as the interactions among them—from networks and their idiosyncrasies.

Not so long ago, conventional wisdom foresaw a world in which desktop computers would operate as autonomous entities, swapping bits with one another over dumb pipes. No need for intelligent networks in that scenario. Each PC would orchestrate all the connections its owner needed, via a kind of oversized Ethernet, one created by government and maintained by good will.

How times have changed, and how much richer our vision of computing and communications has become! Far from an either-or tug of war over whether to place intelligence in the network or on the desktop, today we see the capabilities of networks and their clients as complementary to one another. After all, where can we expect more demanding use of IN-based services, from the owner of a twelve-button telephone, or from one with a powerful multimedia workstation?

Today, we stand at the brink of a new era, one in which users will employ a host of interoperating information appliances—PCs, network computers, set-tops, smart phones, pagers, PDAs, and even "Plain Old" telephones—together with a rich variety of information media, service providers, and network architectures. While much of today's Internet experience will remain available to users who prefer that style of operation, tomorrow's multimedia information appliances will also require the full functionality of traditional telecommunications services. After all, why should we expect users to give up any of the conveniences they have enjoyed in the past, as they add new features to their existing environments?

In particular, while the Internet continues to garner the lion's share of media attention as an application platform, a "one-size-fits-all" network seems unlikely any time soon. Instead, today's networking conferences abound with an alphabet soup of new and refurbished technologies. Solutions galore, each with its own advantage and dedicated adherents. Frames vs. cells, and cells in frames. Gigabit Ethernet IP, of course, and much else besides.

With no single winner in sight, interoperability will offer the engineering community a more realistic template of the future. With the large majority of new data networking equipment devoted to internal information services, today's institutional decisions will surely weigh heavily in shaping tomorrow's merged computing and communications environments.

The dizzying pace of innovation in today's fast-moving world leaves little likelihood of more orderly times ahead. At the same time, however, users have less time and patience for incompatible systems. So much so, that how well a product or service works as a standalone entity counts for far less than how well it works with others.

So what can we expect? Increasingly, integrated information services are provided via cooperating platforms. For instance, a small but growing number of today's "computer users" can enjoy advanced telephony features on their personal workstations employing multimedia call servers linked to corporate PBXs via common signaling. Rather than play telephone tag or wait for someone to read his or her e-mail, a multimedia workstation might establish real-time links to a cellular phone, pager, or wireless modem as part of a conference call, for example. The advent of smarter "peripherals" should spur greater use of IN-based telecommunications.

As information appliances grow more capable and more diverse, so too will the collective demands they place upon the networks that serve them. And herein lies the challenge, which IN designers face: to shield service creation from the unpleasantness of feature interaction and system diversity. In other words, make it all work together, and—while we are at it—make interoperation look easy.

The present volume offers a helpful and detailed guide to implementing this most challenging of network design tasks by providing a detailed compilation of standards, the rationale behind them, and the context in which they operate. With clarity and exhaustive detail, it offers network managers and designers templates for designing good fences—the kind of fences that make good neighbors.

I feel sure that the material contained in these pages will guide and inspire the design of emerging networks as well as the broadband networks yet to come.

ARNO A. PENZIAS
Vice President and Chief Scientist of Bell Laboratories
Lucent Technologies, Murray Hill, New Jersey

Preface

This book was written as the result of several years of the authors' work in IN Standards and their applications to products and services. The effort naturally spans multiple fields and activities, both inside and outside of AT&T, and could have never brought the result had it not been for the assistance of many people.

Many thanks to Arno Penzias, Vice President and Chief Scientist of Bell Laboratories, for his attention to the manuscript, enthusiasm, and willingness to share his vision of the direction of IN in the fore-word to the book.

Hui-Lan Lu carefully reviewed the draft manuscript; her insightful criticism resulted in correcting many oversights and clarifying many points that would otherwise be left obscure. Nilo Mitra and Lev Slutsman helped in shaping the protocol and service creation materials, respectively. Terry Jacobson thoroughly reviewed the manuscript, specifically on the subject of Personal Communications, and provided us with the latest developments regarding Wireless Intelligent Network (WIN). The authors can hardly overestimate the efforts of these experts to whom they are indebted.

The authors wish to acknowledge the help of their Bell Labs colleagues with whom they have been closely working on the subject: Elias Dacloush, Wouter Franx, Tung-Hai Hsiao, Syed Husain, Terry Jacobson, Mark Klerer, Doris Lebovits, Neil Lilly, Hui-Lan Lu, Yota Mastoris, Nilo Mitra, Tim Rinker, Gary Schlanger, Bill Sears, Bill Shores, Lev Slutsman, Cliff Spencer, Mohammad Torabi, Hans van der Veer, Alan Waxman, Huw Weatherhead, Stan Yeh, and Wayne Zeuch.

The knowledge reflected in this book has been gathered through our long association with the representatives of many companies and countries with whom we spent long hours in different parts of the world discussing the very subject described in this book. We would particularly like to thank Evelyn Swenson (Telstra, Australia); Frans Haerens (Alcatel-Bell, Belgium); Nico Weymaere (Belgacom, Belgium); Tom Schaffnit and George Young (MPR, Canada); Joachim Claus (Deutche Telecom, Germany), Norbert Haberer, Ralf Rieken, and Renate Zygan-Maus (Siemens, Germany); Stephan Goerlinger, Jean-Bernard Thieffry, and Bernard Vilain (Alcatel, France); François Gallant, Roberto Kung, and Jacques Muller (France Telecom, France);

Giuseppe Spinelli (Italcable, Italy); Mario Bonatti (Italtel, Italy); Nicola Gatti (Telecom Italia, Italy); Fulvio Faraci (CSELT, Italy); Masanobu Fujioka (KDD, Japan); Sadahiko Kano, Ken-ichi Kitami, Tomoki Omiya, Takeshi Sugiyama, and Masanobu Yoshimi (NTT, Japan); Yun-Chao Hu, Burt Jellema, and Frank Urbanus (Ericsson, Netherlands); Professors Bo Ai and Jun-Liang Chen (Beijing University of Post and Telecommunications, People's Republic of China); Cecilia Ritter (Telia, Sweden); Frank Salm (AG Communications Systems, United States); Harry Hetz (Bell Atlantic, United States); Jim Garrahan, Arkady Grinberg, Deb Guha, Jack Nasielski, Maureen O'Reilly, and Pete Russo (Bellcore, United States); Louis Chong, Jay Hilton, Lev Koyfman, Joe Lenart, Dave Morris, and Tom Walsh (GTE, United States); Norm Kummer, Doug Turner, and John Visser (Nortel, United States); Steve Mecrow, Derek Mainprice, Brian Rae, and Richard Stretch (British Telecom, United Kingdom); and Jane Humphrey and Paul Storey (GPT, United Kingdom).

The authors also want to acknowledge their employer, AT&T Bell Laboratories. Even though this book was not an AT&T project, the superb technical library and computing facilities of Bell Labs were at the authors' disposal.

We are grateful to the book's sponsoring editor at McGraw-Hill, Steve Chapman; editing supervisor, Dave Fogarty; and production supervisor, Suzanne Rapcavage, for their highly professional, cheerful, and clockwork-precise handling of publishing matters.

Finally, it is the high time to thank the International Telecommunications Union for generously granting the authors its authorization as a copyright holder for the reproduction of the ITU-T IN Recommendations material quoted in this book. Of course, the sole responsibility for selecting extracts for reproduction lies with the authors. One goal of this book is to generate an interest in and help understanding the ITU-T IN Recommendations, which can be obtained from

International Telecommunications Union
General Secretariat - Sales and Marketing Services
Place des Nations
CH-1211 GENEVA 20 (Switzerland)
Telephone: +41 22 730 51 11   Telex: 421 000 uit ch
Telegram: ITU GENEVE   Fax: +41 22 730 51 94
X.400: S=Sales; P=itu; A=Arcom; C=ch   Internet: Sales@itu.ch

IGOR FAYNBERG
LAWERENCE R. GABUZDA
MARC P. KAPLAN
NITIN J. SHAH

1

Introduction

1.1 Overview of This Book

The purpose of this book is to help telecommunications professionals understand and use Intelligent Network (IN) Standards, especially those published by the International Telecommunications Union (ITU) in its Q.1200 Series, as well as other closely related standards.

In Chap. 1, the book begins with this overview and proceeds to cover the definition and the history of the IN concept, some background on the services that are driving the implementation and evolution of IN, the standards bodies that are responsible for various aspects of IN standardization, and the structure and a general overview of the ITU IN Recommendations.

Chapters 2 through 6 then introduce the IN Conceptual Model and use it as a framework to methodically describe and explain the ITU-T IN Recommendations, including general Recommendations as well as those specific to the progressively increasing packages of functional capabilities, called *Capability Sets*. To this end, the final (refined) version of Capability Set 1 is described in detail, and the key concepts and results of the ongoing standardization of Capability Set 2 are addressed.[1]

Finally, Chap. 7 completes the story by describing relationships to other key activities and by pointing ahead to the future of IN

---

[1]The reader should bear in mind that until the draft documents of a standard are approved, the material they contain keeps changing—sometimes drastically. For this reason, after consulting the approved Capability Set 2 Recommendations (which are expected to be published by the ITU toward the end of 1997), the reader may detect some differences with what is presented in this book. To minimize the possibility and extent of such differences, the authors have selected the most stable, fundamental material from what is presently in progress.

standardization in terms of work on Capability Set 3, Long Term Architecture (LTA), Telecommunications Information Networking Architecture (TINA), and the related work in regional standards bodies.

1.2 Definition of the Intelligent Network

In our view, the Intelligent Network is an architectural concept that provides for the real-time execution of network services and customer applications in a distributed environment consisting of interconnected computers and switching systems. Also included in the scope of IN are systems and technologies required for the creation and management of services in this distributed environment.

Now, in principle it would be possible in this distributed IN environment to develop each new network service or application in a customized way, to tightly couple the interactions between switches and computers, and to create a specialized IN protocol from the physical layer right on up to the application layer. Indeed, this might be the right approach to developing a highly specialized service, one which required extreme performance, for example. However, fairly early in the history of IN it was realized that the IN would be much more useful as a service-building platform if it adhered to certain principles of independence. These are

Service independence. This means that the IN is indeed a platform architecture intended to support a wide variety of services by using common building blocks.[2]

Clear logical separation of basic switching functions from service and application functions. This principle is essential to IN's promise of speeding up the development of new services, and it does this by making connection establishment available to service processing entities as a cleanly layered function.[3]

---

[2]An interesting question is how we can properly choose the set of building blocks in order to anticipate the needs of future services. The usual approach to this—and the one taken in the ITU standards—is to describe a list of benchmark services whose needs must be satisfied by the building blocks. This does imply, however, that the IN may need to grow when significantly new and different services come over the horizon, by adding new building blocks or stretching old ones. This is pretty much what does happen.

[3]Though simple sounding, this principle is not without subtleties. In fact, some of the greatest controversies in the development of industry standards for IN have swirled around the boundary between service control and connection control. The trick seems to be to make connection establishment available at a high enough level of abstraction so that service processing need not get overly involved in the details of managing connections, which would really defeat the purpose of logically separating the functions in the first place.

Independence of application interactions from lower-level communication details. This principle permits wide flexibility in the physical implementation of INs and emphasizes the important point that the key protocols of the IN are really application layer protocols. [In practice, most IN implementations in today's telecommunications networks use a lower-level protocol suite consisting of ITU-T (formerly CCITT) Signaling System No. 7 and its Transaction Capabilities Part (TCAP) to carry the IN Application Protocol (INAP)]. However, any stack providing similar services would do the job, and this fact will become increasingly important as voice services merge with data services in the unfolding world of Broadband Integrated Services Digital Network (B-ISDN) and Asynchronous Transfer Mode (ATM).

With the simple definition stated previously in mind, along with the three principles, it should now be clear that the main job of industry standards for IN is to specify the interfaces among entities in the distributed, service-independent environment that constitutes the IN. However, before moving immediately into a description of the standards, we will try to add a bit more to comprehension of the IN concept with a short digression into what is by now its rich history in the telecommunications industry.

1.3 Evolution of the Intelligent Network

1.3.1 Long-Distance origins

What we today call the Intelligent Network had its roots in the frustrations of network service planners with the limited, highly specialized capabilities of the switching systems that were available when advanced network services first began to be deployed in the toll (long-distance) network of AT&T. The original architectural solution was designed beginning in the mid-1970s, and systems supporting the first services were deployed by the early 1980s.

At the time when these developments began, computer-controlled electronic switching systems had begun to be widely deployed in the local service networks of what was then the integrated Bell System (Members of Technical Staff, 1986), and these systems were capable of supporting line-oriented services such as Call Waiting and Call Forwarding. Still, the toll network at this time basically had the function of moving large amounts of telephone traffic efficiently among metropolitan areas, and the No. 4A Crossbar switch, which was the backbone of this network, largely implemented simple digit manipulation functions (translating, deleting, and prefixing digits in blocks of three or six) to perform the necessary routing operations. Indeed, the new 4ESS® all-electronic toll switch introduced in 1976, while much larger in capacity, essentially emulated these simple functions with some extensions (Ritchie and Tuomenoksa, 1977, p. 1021).

Nonetheless, the AT&T toll network of the 1970s did implement a few value-added features that were intended to be available on a nationwide basis, and one of these was called Inward Wide Area Telecommunications Service or INWATS. WATS referred to a family of bulk-calling plans offered by the Bell System to business customers, and the Inward version, introduced in 1967, allowed businesses to be reached on a toll-free basis through telephone numbers beginning with the prefix 800.[4] The service was originally conceived as an automated version of collect calling, and enjoyed modest, steady growth through the late 1960s and the 1970s. To implement this service in the network using No. 4A Crossbar switches with their limited capabilities required special routing arrangements, so that the available simple digit manipulation functions could be applied sequentially at Originating and Terminating Screening Offices (OSO and TSO) to screen the call for validity and route it to the proper destination switching office (Sheinbein and Weber, 1982, p. 1738). At this point, a key role was played by the new technology of Common Channel Interoffice Signaling (CCIS) (Ritchie and Menard, 1978). For reasons of efficiency and in order to upgrade the control infrastructure of the network, this system was introduced, also in 1976, and at first was used only to transmit address digits and trunk status information, replacing trunk-associated tone signals. But it was quickly deployed to a subset of the toll switching systems, including the No. 4A Crossbars (which had small computer adjuncts called Electronic Translation Systems) as well as the new 4ESS. Also from the very beginning, it was hypothesized that CCIS, which was really a highly reliable packet data network interconnecting the control processors of the switching systems, would be the basis for new services.

The next step was taken by the Switching Systems Engineering organization of Bell Labs, in particular by Roy Weber and collaborators (Lawser and Sheinbein, 1979), who proposed that a new network element, referred to as a *network database,* be connected to the CCIS network so that, with appropriate addressing, it could be interrogated by any switch that was also on the CCIS network. The initial application envisioned for the network database was a more efficient implementa-

---

[4]The North America Dialing Plan typically reserves the first three digits of a telephone number to a geographic area code. The 800 prefix, which does not denote any area code, was chosen to indicate that the call is supported by a Freephone service (i.e., it is free of charge to the caller). Later other non-geographic-area-code prefixes (e.g, 900 or 500) were adopted to indicate the types of services provided on the calls.

tion of INWATS. Instead of routing the calls via an administrative nightmare of digit manipulations and special code substitutions, the special toll offices handling INWATS, called Originating Screening Offices (OSOs), would temporarily halt call processing and launch a query through CCIS to the network database. The query would contain the dialed 800 number. The database computer would perform a 10-digit[5] translation of this number into a standard routing number, which would be returned to the OSO over CCIS, and the call would proceed. Not only was this more efficient, but it also permitted a company for the first time to have a single national 800 number, since the database was conceived as a nationally available resource. (Indeed, early depictions tended to suggest that it was literally a single computer, though it fairly quickly became clear that capacity considerations would require multiple physical machines, with the database divided among them.)

Following rapidly upon this seminal development, additional enhancements to what came to be called 800 Service were devised, and another team of systems engineers led by Al Mearns of Bell Labs applied the same technology to the automation of Calling Card service (Mearns et al., 1982). Calling Card service was actually implemented first (beginning in 1980) and the deployment of the INWATS architecture (initially called CCIS INWATS) occurred in 1981. On April 25, 1982, the Federal Communications Commission (FCC) approved a tariff for Expanded 800 Service with many new features made possible by the network database implementation.

Even while work proceeded on the initial service applications of the network database, the systems engineering team at Bell Labs realized that this exciting new technology could be even more useful as a general-purpose platform for the development of network-based services (Andrews and Martersteck, 1982). For example, a line of investigation was started to propose "primitive" operations between the switch and network database, which could then be combined to realize different new services. This work culminated in what was known as the Direct Services Dialing Capabilities (DSDC) architecture (Asmuth and Gawrys, 1981; Horing et al., 1982). The name came from the notion that the architecture enabled users to access services by direct dialing, just as the basic network allowed direct dialing of long-distance calls. The architecture included a generalization of the network database which became known as the Network Control Point (NCP), a set of switch functions for access to DSDC which was given the name Action Control Point (ACP), and a set of service-independent commands with

---

[5]The standard size of telephone numbers in North America.

associated parameters which could be exchanged between the ACP and NCP to realize different services. Procedures for service assist and handoff (which later found their way into standards) were also invented at this time. Yet another fundamental component of the IN that was innovated by the Bell Labs group as part of work on DSDC was the Intelligent Peripheral (IP), the first instance of which was called the Network Services Complex (NSCX) and which supported features requiring touch-tone and voice prompt interaction with callers.

The first services implemented using the new DSDC version of the technology were Advanced 800 (which included such features as allocation of calls to different destinations by percentage and customer-alterable routing) and Software-Defined Network (SDN) (Ullrich, 1984; AT&T, 1985). The latter was the first instance of another key IN-supported service: Virtual Private Network (VPN), which allows customers to economically use and manage a portion of the public network for their private network traffic.

The DSDC architecture has been continuously enhanced and still forms the basis for many of AT&T's most successful long-distance services for business and consumer customers. The number of NCPs and IPs of all types has grown beyond the most feverish visions of the early planners, and on a typical business day in 1995 over half of the 200 million or so calls handled in AT&T's network utilized this IN architecture. Furthermore, with the growth of the competitive U.S. long-distance industry, MCI, Sprint, and other carriers evolved their own versions of this technology to support toll-free calling, Virtual Private Network, and other long-distance services.

1.3.2   Extension to local services

In 1984 the court decree breaking up the Bell System was implemented, which had a number of profound effects on the further development of the IN concept. First of all, in the years required to implement the required separation of local and long-distance services, ambitious plans to spread the CCIS network and the DSDC architecture to the local switch level were put on hold. Secondly, the systems engineering team that had developed the architecture was divided, with many key engineers remaining in AT&T Bell Laboratories, where they supported further development of long-distance services, and others going to Bell Communications Research (Bellcore), the new laboratory set up to provide R&D services to the Regional Bell Operating Companies (RBOCs). Finally, a fierce legal battle broke out over the rights to the databases for the fast-growing 800 and Calling Card services. The outcome of this very complicated matter can be summarized by saying that, while it was clear that the RBOCs shared in the patent rights to the DSDC and related technology developed before the breakup, they were going to have to eventually deploy their own physical databases to support local versions of the services.

Bellcore accordingly went to work developing requirements for a local services version of the network database architecture, and around this time began heavily promoting the term Intelligent Network to describe it (Hass and Robrock, 1986).

With the RBOCs operating as fully independent businesses, some new business needs influenced the IN requirements being put together by Bellcore. One of these was multivendor supply. Northern Telecom (now Nortel) had begun to sell substantial numbers of switches to the RBOCs even before divestiture, but this trend now accelerated greatly and other vendors were added as well to the base of AT&T Network Systems' 1AESS® and 5ESS® switches. Clearly, the new local IN had to not only support but also encourage a multivendor environment. Related to this was the impact of one key provision of the court decree, the one which prohibited the RBOCs or their Bellcore laboratory from manufacturing equipment, including switches. Service planners who were now working at the RBOCs and Bellcore had been accustomed to developing requirements which were eventually translated by Western Electric into switch features through business processes in the unified Bell System. Now this process required sometimes frustrating negotiation with multiple vendors. The IN appeared to potentially speed up this process and make it less frustrating by allowing service functionality to be migrated from the switch to the network database element, now called the Service Control Point (SCP), where it could be maintained and extended by Bellcore or RBOC software developers (software development was permitted by the court decree).

In the mid-1980s, Bellcore accordingly developed specifications for an advanced version of the IN, which it called IN2 (giving the name IN1 to its first local IN specifications for relatively simple services like 800 and Calling Card). This version proposed many advanced capabilities, such as triggers[6] that could occur in the middle of a call[7] and the manipulation of connection legs (in multiparty calls) by the SCP through commands like *create, split, join,* and *free.* Some of the original DSDC architects who had gone to Bellcore contributed to IN2, and so had the rare privilege of doing something over again with insights gained the first time. While it was undoubtedly a very sound piece of work intellectually, vendors considered the required developments to be daunting and it was never realized in this form at the time. With this experience,

---

[6]Triggers are call-related conditions that cause a switch to interrupt its own processing of the call and launch a query to an SCP. The occurrence of the "800" string in place of an area code is an example of a trigger.

[7]For example, when a party flashes the hook in the middle of a call.

Bellcore then moved to even more directly involve vendors in the next round of planning, through the so-called Multivendor Interactions Forum. Around this time another trend made itself felt, and this was the increasing tendency of the RBOCs to develop independent business and technical directions. Several RBOCs, notably U.S. West, Ameritech, and BellSouth, developed their own views on IN architecture, in many cases supported by extensive experiments and work with vendors. Bellcore then worked to incorporate these views into its overall requirements.

The main result of this phase of Bellcore IN work was a specification designated Advanced Intelligent Network (AIN) Release 1.0. Unfortunately, it essentially met the same fate as the earlier IN2: vendor agreement for implementing the total breadth of this specification could not be obtained. Bellcore's response was to define a sequence of smaller-scale releases designated AIN 0.0, AIN 0.1, etc., and this essentially remains the trajectory of the Bellcore requirements work up to the present.

Clearly, the deployment of the IN concept in the U.S. local services market has been a story of some frustration, but it may be worthwhile to step back and observe just how ambitious IN had become by this time (circa 1990) and how many agendas it had taken on. A concept originally designed to build new long-distance services based on flexible routing, screening, and billing capabilities was now being expanded to support line-oriented services, with their potential for interaction with the rich base of line features in local switches, to devolve connection handling from those same switches, and to operate in and foster multivendor environments.

The good news is that at this writing a number of Local Exchange Carriers (LECs) have fought past these obstacles (and others too numerous to detail in this short account) and are deploying innovative IN-supported services to a significant customer base. Further, it is becoming increasingly clear that IN technology holds the key to capabilities that will be required in the fast-approaching era of local services competition. For example, IN's tested ability to perform high volumes of number translations underlies most current proposals for local number portability.

1.3.3   The international Intelligent Network

During its first decade, IN was a largely U.S. phenomenon. This was due mostly to the early advent of long-distance competition in the United States, which drove aggressive deployment of common channel signaling and NCP-based services (as well as to the unique situation of the RBOCs described above). However, by the late 1980s Signaling System No. 7 was an international standard and a well-known technology in many countries, and the first inklings of competition helped spark world interest in services such as Freephone, which had been initially deployed in several countries on a switch-based implementation (just as was the original 800 Service in the United States). Other services, such as VPN and Calling Card, by their nature demanded international implementation. Another driver was the effort by North American vendors to capitalize on their early experience with the development of IN-based network equipment. For example, AT&T Network Systems developed an IN product line for international sale based on the DSDC implementations it had built for the AT&T network, and an instance of this was installed in the British Telecom network to support Freephone by 1988. This was followed by installations in Spain and Italy (Workman et al., 1991). In Japan, NTT had implemented free-dial service on a Network Service Control Point (NSP) element as early as 1985, and introduced a service-independent version of the NSP by 1990 (Suzuki, 1993). France Telecom began working on an IN in 1986 and had implemented IN capabilities in conjunction with the Itineris mobile network by 1991 (Kung and Paul, 1995).

In 1989 a standards project was initiated in the CCITT (now ITU-T) to develop Recommendations for IN interfaces and protocols. Many vendors worldwide then began developing SSP and SCP products and many operators began to deploy them as international standardization activity proceeded.

1.3.4 Things to come

So, if this is the somewhat convoluted past history of the IN, what is its future? One line of thinking is that there will be no future, because the very notion of an "Intelligent Network" is a profound anachronism in a world dominated by the Internet paradigm, which in a popular interpretation consists of a simple network with highly intelligent endpoints. This view is undoubtedly oversimplified. A view widely held by the research organizations of large carriers, and being elaborated upon by an international consortium called Telecommunications Information Network Architecture (TINA-C) (Barr et al., 1993), is that today's IN will evolve to a more general-purpose Distributed Processing Environment supporting client/server architecture, with sophisticated services being offered over it by network providers, end users, and numerous third parties. Actually, this appears to be a natural step (or series of steps) and, though founded upon more than a decade's advances in the understanding of distributed computing, is really remarkably consistent with the original vision of the IN as a distributed infrastructure allowing specialized computers and switches to cooperate in the provision of services (Andrews and Martersteck, 1982, p. 1577). The key to successful evolution is to bring the user, now equipped in many cases with powerful personal computer hardware and software, fully into the picture.

1.4   Intelligent Network-Based Services

Since the modern IN is a service-independent platform for the realization of services, asking the question, "What services may be supported by IN?" is very much like asking, "What applications may be programmed on a computer?" But, just as certain key applications have driven the computer industry through important evolution steps (e.g., accounting and scientific applications supporting the growth of mainframes, spreadsheets and word processing accelerating the PC revolution, consumer and business multimedia applications driving the industry today), the evolution of the IN has been (and continues to be) influenced by certain paradigmatic service demands. This section will review some of the "classic" IN applications which are still big revenue generators, and then discuss the new services that are driving the evolution of the IN.

1.4.1   Classics

As discussed in the previous section, the applications which stimulated the early IN work in the United States were (to give them their generic industry names) Freephone, Credit Card Calling, and Virtual Private Network (VPN). All three of these services (and their close relatives), as implemented by U.S. long-distance carriers, experienced strong growth and had major impacts on these companies. They even changed the way that businesses did business and the way that people conducted personal activities such as shopping.

The growth of 800 Service in North America following the move to IN implementation was nothing short of astounding. With the greater convenience, more flexible features, and lower prices enabled by IN implementation, whole new industries based on concepts like catalog shopping were created, and even quite small organizations and businesses became 800 number subscribers.

Figure 1.1 illustrates one of the simple but powerful features which added to the customer appeal of 800 Service in its IN implementation: Time and Day Routing. As shown, a company might wish to direct calls to its 800 number to an East Coast location during normal working hours in that time zone, to a central states location during a different (overlapping) set of hours, and to a 24-hour location on the Pacific Coast at other times. Another company may wish for its business purposes to route calls to one center Monday through Thursday, and to a second center on Fridays and weekends. Using the methods of *call allocation,* the percentage of calls sent to each center can also be varied.

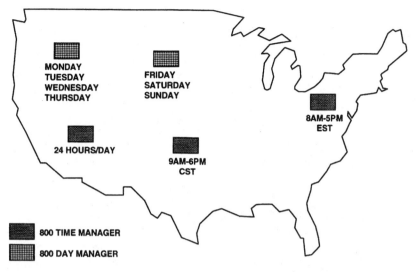

Figure 1.1 Time and Day Routing.

These routing instructions are all contained in customer records stored in the NCP. Furthermore, these records can be created and changed by the customers themselves and downloaded into the network for execution—an instance of service customization!

AT&T's introduction of Software Defined Network (SDN), and competing services from MCI, Sprint, et al., had a transformative effect on the way businesses in the United States put together their internal networks, marked by a sharp trend away from the use of physical private lines and the equipment (multiplexers, etc.) needed to manage private line networks. Figure 1.2 illustrates the components of the SDN. The SDN call works as follows. The calling party dials a number which can either be a standard number from the dialing plan or a special private number. The ACP (switch) recognizes the call as SDN-based on the calling party number (ANI) or trunk number in the case of a directly connected PBX, and launches a query to the NCP. The NCP uses the information in the query to access the appropriate customer record, screens the call, performs a number translation (if necessary), and replies to the ACP with routing and billing instructions. Many optional features can be added to suit the needs of particular industries or customers for private networks. For example, the NCP can request attachment of an Intelligent Peripheral to collect an authorization code to permit the completion of certain call types or to collect an account code for the customer's convenience in charging calls to particular accounts.

Many other services have been implemented on the standard IN architecture and are earning revenue for carriers around the world—

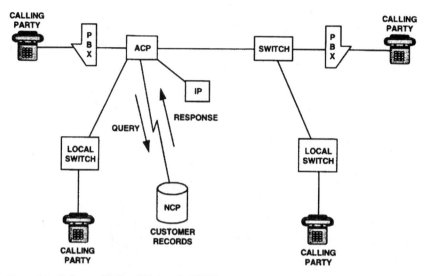

Figure 1.2 Software Defined Network (SDN).

already "classics" or fast becoming so. Calling Card services and Multiquest™-type information services account for large traffic volumes. In the local networks, Network Redirect, in which calls are sent to an alternate business location designated by prestored customer records in the event of disaster is a burgeoning service. IN implementations have also been used to build newly popular Voice Dialing services, and to add richer features to such standard line-side services as Call Forwarding.

1.4.2 Wireless and personal communications services

One of the simple tricks that IN performs so well (known in telephony terms as "number translation") is the mapping of one "name" into another. This may be the mapping of an 800 number into a routing number (Freephone), or a private network number into a public number (VPN). In the world of wireless and personal communications services, this becomes the crucial function of mapping a moving customer's identity into a current location in physical network space. The importance of interconnectivity between service networks is especially important to mobile communications systems.

Providing personal communications service within a wired network is to some degree a straightforward extension of basic IN capabilities. For example, AT&T currently offers a personal number service called AT&T True Connections in which each customer is assigned a personal, lifetime phone number beginning with the prefix 500. When a call to one of these numbers enters the network, a query is launched to a centralized database where the 500 number is translated to a current routing number. When the customer moves to a different location (and wants calls to follow), he or she can call the network to update the database or register for other options such as voice messaging.

As far as the wireless telephony is concerned, even if IN had not been conceived, the need for different service providers in different geographic regions to have common signaling, database, billing, and call delivery technology is immediately apparent for a cellular communications system. Without these links, a cellular customer who roams outside the immediate serving area of his or her service provider would not be able to gain access to a cellular service (or—in some instances— would be able to access the service, but the service provider would not have the information for the user to be billed!). The principles of IN were rapidly adopted by the cellular telecommunications industry during the creation of the standards for the networking protocols and network connectivity for cellular and mobile systems.

The initial work on cellular networks was driven quickly to utilize the work on AIN, but it also had a different focus from the conventional IN trajectory. Whereas the focus of the IN standardization work was on value-added services in the fixed networks, the work on wireless networks was to assist service providers and end users in truly achieving the promise of seamless roaming across geographic boundaries and ubiquitous service networks, which was the premise of the cellular concept.

The Wireless Intelligent Network (WIN) concepts have been developed based on the rapid emergence of cellular and PCS networks over the past decade. The basic requirement of a mobile network is to provide its users with the ability to initiate and receive calls regardless of their location. In the early days of cellular systems, within the confines of an isolated single service provider network, with all the equipment provided by one manufacturer within the service area of the service provider, such a requirement was met. However, as cellular roaming crossed state and country boundaries, the technical and business arrangements to allow initiation and termination of calls across national and international boundaries became a natural expectation of the mobile customers.

From the service provider perspective, such capabilities also allowed the use of IN not only to provide point-to-point telephony, but also to incorporate capabilities for rapid introduction of new services and customization of such services according to the subscriber needs, thus embracing the original intent of the IN. The WIN architecture provides a framework for interruption of call processing at triggers, and querying databases to determine the treatment of the call, depending on the provisions made by the subscriber and the service provider. The WIN architecture is structured so that the triggers and signaling can be made independent of specific services, so that the services can be constructed using external service logic.

An essential WIN requirement is that the IN features and services coexist with conventional switch-based service features (such as Call Waiting). Moreover, to the end user such services must appear to be transparent, regardless of the origin of the service, either from the switch or WIN. Further, the WIN emphasizes open interfaces, so that the end user can roam across service provider networks that may have been integrated by different equipment providers but interoperate to provide transparency of service capabilities. These capabilities are determined by the special agreements between the service providers. This level of interoperability and transparency has led to significant efforts in the standardization of INs for wireless systems, and are embodied in the IS-41 set of standards published by the Telecommunications Industry Association (TIA).

The Mobile Station, Base Station, Mobile Switching Center, Authentication Center, Home Location Register (HLR), and Visitor Location Register (VLR) (see Fig. 1.3) are conventional elements of the cellular and PCS wireless access networks. A Mobile Station is the interface equipment used to terminate the radio path at the user side. It is the Mobile Station that provides a user with the capabilities to access network services. The authentication information related to a particular Mobile Station is managed by an Authentication Center.

The Mobile Switching Center is a Service Switching Point (SSP) for wireless networks, and it is the point in the network that detects the IN triggers. The Mobile Switching Center also constitutes the interface for user traffic between the wireless network and other public switched networks, as well as to other Mobile Switching Centers.

A subscriber's identity is assigned to an HLR, which keeps the subscriber's (and his or her Mobile Station) information and provides service control and mobility management for one or more Mobile Switching Centers on behalf of the subscriber. An HLR may be located within the Mobile Switching Center (and indistinguishable from it); it may also be located within a Service Control Point.

A VLR is used by a Mobile Switching Center to retrieve information for handling calls to (or from) a visiting user. A VLR provides mobility management for one or more Mobile Switching Centers on behalf of a subscriber in visited networks. Similarly to an HLR, a VLR may be located within the Mobile Switching Center (and be indistinguishable from it).

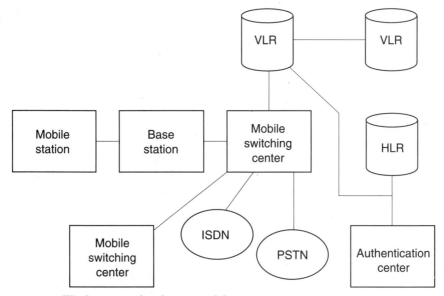

Figure 1.3 Wireless network reference model.

At the moment of this publication, the cellular industry is completing the specification of WIN (Cellular Telecommunications Industry Association, 1996), which builds on the call models of the ITU-T IN standards and adds special functions needed by wireless networks. In the United States, this work is performed by the Telecommunications Industry Association (TIA), an accredited standards body. The WIN work is performed by the Subcommittee TR-45.2, Intersystem Operations. Specifically, WIN enhancements are planned for the next revision of the IS-41 standard, IS-41-D, which should be completed in the second half of 1996.

1.4.3 Multimedia and broadband services

In the arena where telecommunications, computing, and entertainment converge, new multimedia and broadband services are being defined. Some will be straightforward extensions of familiar IN-based services (broadband 800 Service, perhaps), and their implementations may be similarly straightforward. Others present genuinely challenging issues for the network services designer, and are the subject of much current research.

One example may be the class of services involving multipoint, multimedia conferencing. Such services appear to require possibly frequent requests to add and drop conference legs, adjust available bandwidth, and change available quality of service parameters. If the network is to provide added value in these interactions, it must be aware of and able to respond to requests. In the IN context, this partly involves the classic question of the interface between service control and connection control. The reader may recall from Sec. 1.2 that the key principle to keeping this interface manageable is to choose an appropriate level of abstraction—a way for service control to call on the capabilities of connection control without getting tangled up in its complexities and stringent real-time concerns. One proposed approach to doing this is a methodology referred to as *Call Configurations* (Isidoro, 1992). Applying this approach to multimedia IN is an active subject of research and discussion in standards bodies.

1.5   Relevant Standards Bodies

The development of the international IN standards is taking place in the International Telecommunication Union (ITU), particularly in its Telecommunication Standardization Sector (ITU-T) (formerly CCITT).[8] The standards are developed based on contributions brought on behalf of the companies and countries that are members of ITU. The IN standards have been developed in a truly international environment, through active participation of delegations from Europe, the Pacific Rim, and the Americas. At the scheduled ITU-T meetings, each representative is part of his or her country's delegation.

To ensure the agreement among the members of a delegation, many countries have established the bodies that approve (or disapprove) the contributions proposed by the companies and establish the agreed positions which must be upheld by all members of a country's delegation at a meeting. Conversely, once the ITU-T Recommendations[9] are written and approved, the same national or regional bodies that contributed to their development are often accredited to adopt them as standards.

While most of this book focuses on the ITU-T Recommendations, in the next few paragraphs we briefly review the work of the regional bod-

---

[8]ITU-T is a relatively new (circa 1993) name for the organization that was formerly called International Telephone and Telegraph Consultative Committee and well known by its French acronym CCITT. ITU is an arm of the United Nations; in addition to ITU-T it has two more organizations, ITU Radio Sector (ITU-R) and ITU Development Sector (ITU-D). The IN work was initially carried out in CCITT Study Group XI (Switching and Signaling) and Study Group XVIII (ISDN). After reorganization of ITU in 1993, this work has been carried out entirely in Study Group 11 (particularly, in its Working Party 4).

[9]Although the ITU-T Recommendations are referred to—even in this book—as standards, they are called "Recommendations" to indicate that it is up to a given country or a region to make the recommended material become a standard.

ies in the United States, Europe, and Japan.[10] In this chapter we address that part of the work which deals with the preparation and approval of the contributions to ITU-T. In Chap. 7 we return to this subject in order to cover the development of regional standards for IN.[11]

The United States is formally represented in ITU-T by the U.S. Department of State. All contributions to the official ITU-T meetings are to be formally approved by the U.S. International Telecommunications Advisory Committee (US ITAC). All IN contributions that are brought to the US ITAC are first approved by the committee T1S1, which is an arm of the Committee T1 sponsored by the Exchange Carriers Standards Association (ECSA) and accredited by the American National Standards Institute (ANSI). Within T1S1, whose responsibilities include architecture and signaling standards, the IN Sub-Working Group (SWG) created within the sub-committee T1S1.1 has been reviewing the IN contributions.

This process not only ensures that the representatives of the U.S. companies reflect common views at the international meetings, but it also fosters early industry agreements that may lead to the formation of the national standards. Indeed, the IN SWG, whose initial task was to prepare the U.S. positions to the ITU-T meetings, later was entrusted with the first ANSI IN Standards project. (The status of this project is discussed in Chap. 7.)

The European Telecommunications Standards Institute (ETSI) is governed by the Technical Assembly, to which Technical Committees report. Two such Technical Committees, Network Architecture (NA) and Signalling Protocols and Switching (SPS) (or, more precisely, their respective subcommittees, NA6 and SPS3), have been involved in both the preparation of European positions to ITU-T and development of the European IN standards.

In Japan, the consensus-building body is the Telecommunications Technology Committee (TTC), which is also responsible for the development of national standards. The IN work is performed in the IN Working Group, which reports to the Technical Subcommittee I. Technical subcommittees report to the Technical Assembly governed by

---

[10]Note, however, that the participation in and contribution to the work on the IN standards has not been limited only to the United States, Europe, and Japan, although their respective delegations were, and still are, the largest. From the onset of IN activities, delegates from Canada and Australia have been among the strongest contributors. And as the IN has grown around the world, Argentina, Chile, China, Korea, India, Indonesia, and Singapore have actively joined the work on the IN standards.

[11]Hetz and Rinker (1995) contains a comprehensive overview of the standards processes in different bodies. For the overview of the early years of IN standardization see Visser (1991).

the Board of Directors, which, in turn, reports to the General Assembly of TTC. The Ministry of Post and Telecommunications of Japan formally represents that country in ITU-T; the technical work is done by the representatives of Japanese companies, whose contributions have been approved by TTC.

1.6   The Structure and Contents of the ITU-T IN Recommendations

The Q.12xy Series of ITU-T Recommendations has been assigned a block of 100 numbers. The digit y indicates the topic of a Recommendation, while the digit x indicates whether the Recommendation is *general* (in which case, $x = 0$[12]) or whether it belongs to a specific *Capability Set* (CS) (in which case, x is the number of such a CS). The exception is Recommendation Q.1290, "Glossary of terms used in the definition of Intelligent Network Recommendations," which covers the terms common to all existing CSs.

In March 1993, the World Telecommunications Standardization Conference formally approved the first set of the series, which contained the general Recommendations as well as those of Capability Set 1 (CS-1).[13] Soon after that, the work on the CS-1 *refinements* started, which was completed in May 1995.

The work on the CS-2 Recommendations is still in progress; it is presently scheduled to be completed no later than in the beginning of 1997. The work on CS-3 has started recently. This book reports on the current CS-2 and CS-3 work; however, the only definitive international IN standard as of this writing (March 1996) is CS-1R.

The summary of the existing IN Recommendations is as follows:

- Recommendation Q.1200, "General Series Intelligent Network Recommendation Structure," explains the naming conventions and provides the outline of the Q.120x Series.

- Recommendation Q.1201, "Principles of Intelligent Network Architecture," defines the objectives and provides the overall description of IN. In addition, this Recommendation contains high-

---

[12]Thus, Recommendations Q.1200, Q.1201, Q.1202, Q.1203. Q.1204, Q.1205, and Q.1208 form the general Series. Recommendations Q.1201, Q.1202, and Q.1203 were developed jointly by Study Groups 11 and 18, for which reason they have been assigned additional I-Series names, which are I.312, I.328 and I.329, respectively. The rest of the Recommendations have been developed by Study Group 11 alone.

[13]Except for Recommendation Q.1219, which was completed in May 1993 and approved in April 1994.

level IN functional requirements, and it describes the IN architectural concept. To this end, the Recommendation presents the IN Conceptual Model, which is an architecture based on four *planes: Service Plane, Global Functional Plane* (GFP), *Distributed Functional Plane* (DFP), and *Physical Plane.* Each of the rest of the IN Recommendations (except Recommendation Q.1290) deals with an architecture related to one of the above four planes.

- Recommendation Q.1202, "Intelligent Network—Service Plane Architecture," describes the IN Service Plane, making a point that all IN-supported services can be described to the end user or subscriber by means of a set of generic blocks called *Service Features.*

- Recommendation Q.1203, "Intelligent Network—Global Functional Plane," describes the architecture of the IN Global Functional Plane. To do so, it introduces the concept of Service-Independent Building Blocks (SIBs), which are modeling constructs that denote the network-wide capabilities needed to deliver Service Features.

- Recommendation Q.1204, "Intelligent Network—Distributed Functional Plane," defines the IN architecture in terms of IN *Functional Entities* (FEs), which are sets of functions that reside in a single piece of physical equipment. The FEs execute *Functional Entity Actions* (FEAs) and communicate with each other by exchanging *Information Flows* (IFs) over (logical) media called *relationships.* The SIBs are realized in the Distributed Functional Plane through the distributed processing carried by the FEs. The Recommendation also defines the IN call modeling concept and provides a general example of the *Basic Call State Model* (BCSM), in which the IN *triggering* concept is defined.

- Recommendation Q.1205, "Intelligent Network—Physical Plane," defines the architecture of IN in terms of *Physical Entities* (PEs), which constitute the IN equipment, and their interconnections.

- Recommendation Q.1208, "General Aspects of the Intelligent Network Application Protocol (INAP)," specifies the methodology for the development of INAP, its main point being that the design of the protocol must be based on the standardized Open Systems Interconnection (OSI) Application Layer principles, which are technically sound and have been verified and supported by the computing industry for years.

- Recommendation Q.1210, "Q-Series Intelligent Network Recommendation Structure," provides the outline for the whole series of CS-1 Recommendations.

- Recommendation Q.1211, "Introduction to Intelligent Network Capability Set 1," specifies the contents of CS-1 and defines its service- and network-related principles. Two key principles of CS-1[14] are *Single-Endedness* and *Single-Point-of-Control,* which, respectively, state that IN service logic may directly affect only one half-call (either originating or terminating) and that only one instance of service logic (i.e., service logic process) can be in contact with any half-call.

- Recommendation Q.1213, "Intelligent Network—Global Functional Plane for CS-1," specifies 14 CS-1 SIBs.

- Recommendation Q.1214, "Intelligent Network—Distributed Functional Plane for CS-1," forms the basis for the definition of the CS-1 INAP. The Recommendation defines CS-1 FEs and provides their models related to service execution. In particular, the Recommendation defines the CS-1 BCSM. Recommendation Q.1214 also describes each of the CS-1 SIBs in terms of FEAs performed by all involved FEs and the IFs exchanged among them on behalf of those SIBs. Finally the Recommendation supplies detailed description of all CS-1 IFs.

- Recommendation Q.1215, "Intelligent Network—Physical Plane for CS-1," lists the IN PEs and describes the allocation of IN FEs to these PEs as well as the involved interfaces.

- Recommendation Q.1218, "Intelligent Network Application Protocol," specifies the protocol to support the capabilities required by the CS-1 benchmark services (defined in Recommendation Q.1211).

- Recommendation Q.1219, "Intelligent Network User's Guide for Capability Set 1 (CS-1)," reflects invaluable clarifications of many CS-1 issues, including identification of specific problems with proposals for their solution. In addition it provides systematic description of several service scenarios [most notably, the Universal Personal Telecommunication (UPT) service scenario].

- Recommendation Q.1290, "Glossary of Terms Used in the Definition of Intelligent Network Recommendations," defines IN terms and concepts and contains the list of acronyms, which makes it a necessary companion when studying other Recommendations.

The structure of the CS-2 Recommendations is parallel to those of CS-1. The CS-2 Recommendations to be issued are Q.1220, Q.1221, Q.1222, Q.1223, Q.1224, Q.1225, Q.1228, and Q.1229. In addition, all general IN Recommendations and Recommendation Q.1290 will be revised and reissued together with the CS-2 series.

---

[14]These principles remain valid in CS-2.

2

Principles of IN

2.1 Overview

The trends in telecommunications history outlined in the previous chapter illustrate both the necessity and economic feasibility of distributing the call and service software in the network in order to bring new services to the market faster and cheaper. What used to be part of a switch's software can be developed independently on a general-purpose computer, and the services are realized via communications (traditionally called *signaling*) between general-purpose computers, switches, and specialized devices.[15] The word "intelligent" then applies to networks interconnecting such equipment, with the understanding that the role of these networks is to support the distribution of the "intelligence" and that the signaling within the networks is *service-independent*. To stress the importance of the latter requirement, we note that distributed computer programs are much more expensive to develop (and such development is inherently long) than nondistributed ones. For this reason, service independence is the very essence of IN: an IN network must hide software distribution from service programmers. *Network independence* is as important: a service ought to change very little—or, better, not change at all—when the network structure changes. There is an interesting manifestation of duality in these principles: the former postulates that the signaling structure of the

---

[15]These devices can play announcements, collect dialed digits, recognize voice, provide bridging facilities for conference calls, etc. Before the introduction of IN, such functions could have been performed only by the switches. In effect, the emergence of such devices demonstrates the migration from the switch and distribution over the networks of *hardware* functions, vis-à-vis non-switch-based service development, which manifested the migration and distribution of *software*.

network is invariant relative to the services; the latter postulates that the service development platform is invariant relative to the network equipment and its interconnection.

As far as ITU-T IN recommendations are concerned, there is only one, Recommendation Q.1201, entirely dedicated to addressing IN principles. To this end, Recommendation Q.1201 deals with definitions, requirements, methodology, and architectural concepts that are applicable to both present and future technologies (and relevant standards). This recommendation is open-ended in that some capabilities and interfaces discussed in Q.1201 are standardized in CS-1, others are left to later Capability Sets, and yet others—specifically the ones presently experimented with in research laboratories—are the subject of study by the Long Term Architecture (LTA) group rather than explicit standardization.

The rest of this chapter is almost entirely dedicated to Recommendation Q.1201, which culminates in the introduction of the Intelligent Network Conceptual Model (INCM).The INCM, however, may appear too abstract to the reader, for which reason we added a section that explains it by systematically drawing analogies from well-understood computing concepts.[16] Finally, there is a section that explains the role of the Q.12x1 recommendations and their place in this book.

2.2   Recommendation Q.1201, Principles of Intelligent Network Architecture

2.2.1   Summary

Recommendation Q.1201 is 33 pages long. The first (and still present) publication of the recommendation is dated October 1992. This recommendation was developed together with ITU-T Study Group 18, which assigned it its second name, I.312.

Recommendation Q.1201 gives a comprehensive description of IN standardization as far as its motivation, objective, definition, scope, plans, and relation to other fields in telecommunications and computing are concerned. Recommendation Q.1201 also provides IN requirements and introduces the IN Conceptual Model (INCM).

In what follows, we review the recommendation material on

1. Motivation, objectives, scope, and overall principles of IN

2. IN functional requirements

3. IN architectural concept

---

[16]This material is based on Faynberg et al. (1993).

2.2.2 Motivation, objectives, scope, and overall principles

Recommendation Q.1201 explains that the term *Intelligent Network* is used "to describe an architectural concept which is intended to be applicable to all telecommunications networks." This and the rest of the IN Recommendations are motivated by the "interests of telecommunications services providers to rapidly, cost-effectively, and differentially satisfy their existing and potential market needs for services." At the same time, "these service providers seek to improve the quality and reduce the cost of network service operations and management." Recommendation Q.1201 cites examples of technologies that have enabled IN: mobility and digital transmission (both enabled in turn by semiconductor technologies), signaling (i.e., data communications protocols), and distributed databases.

According to Recommendation Q.1201, the objective of IN is "to allow the inclusion of additional capabilities to facilitate provisioning of service[s], independent of the service/network implementation in a multivendor environment." [This formula explains the role of equipment vendors in IN standardization: to ensure that the demands of service providers can be met in the planned equipment; and if certain demands cannot be met (either technically or economically), it is for the equipment vendors to articulate that.] The scope of IN includes virtually all types of networks: Public Switched Telephone Network (PSTN), Packet-Switched Public Data Networks (PSPDN), mobile networks, and both Narrowband and Broadband Integrated Services Digital Networks (N-ISDN and B-ISDN).[17]

In defining IN as "an architectural concept for the operation and provision of new services," Recommendation Q.1201 lists several specific factors that both illustrate the definition and place essential requirements on the IN architecture. One such requirement is that the communication interfaces between various types of IN equipment be standardized; another is that service creation and management be standardized and both service subscribers and service users gain runtime access to certain service attributes. While the importance of the former requirement is obvious, the latter one (an example of which was demonstrated in Chap. 1) is worthy of further discussion. First, we should consider a few definitions. With respect to telecommunications networks, an organization that owns a network is called a *network*

---

[17]It may be worth glancing ahead into the next chapter at this point. As far as the actual standardization goes, the Capability Set 1, which is the first set of IN Recommendations, has mainly concentrated on *fixed* or *wireline* (versus mobile or wireless) N-ISDN networks, while CS-2 and CS-3 have included the support of both mobility and B-ISDN services.

provider (or carrier). With the introduction of IN, a network provider may be different from a *service provider,* which is the organization that actually owns the service. Effectively, service providers *run* the services in the networks that belong to network providers similarly to the way data-processing centers run their programs on the off-premises shared-time computers. (We will further explore this analogy later in this chapter; for now, it is important to observe that the networks and the services provided through them are being *separated* both functionally and in terms of their respective ownership.) A *service subscriber* is an organization (or a person) that purchases the service, and a *service user* is a person who accesses the service.[18] The IN definition postulates that service providers should have a standard (and network-independent) mechanism for service programming and its introduction into the network, while service subscribers and service users should be able to customize certain aspects of services. The very fact that these capabilities have been considered by the standards bodies reflects the process of deregulation that is dominant in the telecommunications industry today.[19] Naturally IN is becoming the very vehicle of decoupling network provision from service provision.

2.2.3   IN functional requirements

Recommendation Q.1201 divides functional requirements into two parts: service requirements and network requirements. Both apply to the five areas defined as follows:

1. *"Service Creation.*   An activity whereby supplementary services are brought into being through the specification phase, development phase, and verification phase."

2. *"Service Management.*   An activity to support the proper operation of a service and the administration of information relating to the user/customer and/or network."[20]

3. *"Network Management.*   An activity to support the proper operation of an IN-structured network."

---

[18]Using the example of the Freephone service (better known as 800 service in North America) explored in Chap. 1, a service subscriber is the entity that owns the Freephone number, while the people who actually dial (or answer) Freephone numbers are service users.

[19]The process originated in the United States in 1986 when Federal Communications Commission issued a mandate for an Open Network Architecture (ONA) for equal access to basic telecommunications services with the goal of ensuring that competitive service providers can develop enhanced services in the most economical way. Later, similar processes took off in Europe and Japan. Presently, there is a subgroup within the IN group in ITU-T Study Group 11 that meets to discuss the ONA issues at every meeting.

[20]The "or" in this definition is nonexclusive.

4. *"Service Processing.*   Execution of network functions in a coordinated way,...[so] that basic and supplementary services are provided to the customers."

5. *"Network Interworking.*   A process through which several networks (IN-to-IN or IN-to-non-IN) cooperate to provide a service."

The overall requirements stipulate, among other things, that both *Plain Old Telephone Service* (POTS) and ISDN interfaces provide access to services in IN, that the services may span several networks, that multiparty services be supported, and that IN networks be responsible for collecting the service use information (e.g., charging and performance statistics). Recommendation Q.1201 points out that IN—from the end-user point of view—is transparent in that "no service processing requirements can be identified that have specific reference to the IN as such." Instead, IN networks must support a set of service and access capabilities. The service capabilities listed in Recommendation Q.1201 include virtually all bearer services, teleservices, and broadband interactive and distribution services; the list of the access capabilities is again all-inclusive, covering fixed public and private network access, mobile network access, and broadband network[21] access. The catchall nature of the list of capabilities to be supported by IN, in this case, does not render the requirements too general; it rather stresses the main point regarding service delivery in IN: it is network-independent (and, consequently, network-hardware-independent). To continue our analogy with computers, we observe that an application program written in, say, the C language, in general produces the same result whether one runs it on a personal computer or a supercomputer. Furthermore, the type of monitor (or terminal) in front of the computer user is irrelevant (as long as such a monitor meets certain standards) to the program as is the proximity of the monitor to the computer or the access method employed to connect the monitor to the computer. Similarly, a service designed for an IN network should produce the same result whether the network is POTS or ISDN, and the mode of the access to the network should have no effect on the service.

Clause 2.2.5.2 of Recommendation Q.1201 presents the IN service processing model in Fig. 16, which is reproduced here in Fig. 2.1.

The circles in this figure represent the switching nodes. If we disregard momentarily the part of the figure above the circles, we will be back in the non-IN world, in which the switching nodes cooperate in delivering service to the parties involved in a call. The services may be basic or supplementary, but they are programmed within the switching nodes (that is, the *service logic* is contained within the switching

---

[21]Both Asynchronous Transfer Mode (ATM) and Synchronous Transfer Mode (STM).

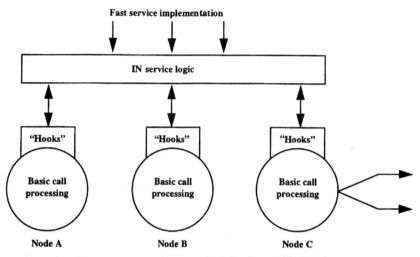

Figure 2.1   The IN service processing model. *(After Fig. 16/Q.1201.)*

nodes). In the IN world, however, the service logic for the supplementary services *may* be developed and executed outside of the switching nodes. The latter run cooperatively the *Basic Call Process* (BCP),which provides basic services and (when supplementary services are requested) may trigger the execution of the IN service logic by means of software "hooks."[22] The hooks and the messages between the switching nodes and the rest of the IN equipment are the subject of IN standardization.

This model, at the present level of abstraction, is central to IN. Through step-by-step refinement of the definitions and functions of its components, concrete specifications for each IN Capability Set are derived. We will return to the model in the next section, but to finish the present discussion, we address two additional important subjects covered (although intentionally left open-ended) in Recommendation Q.1201. These subjects are network interworking requirements and Long Term Architecture (LTA) framework.

The recommendation addresses the network interworking requirements on a rather high level of abstraction: it just states that *gateway functions* may be used for internetwork access for (a) Service Processing, (b) Service Management, and (c) Service Creation; and names these gateway functions:

---

[22]It is essential that *only* supplementary services (not the basic ones) are to be provided outside of switching nodes. There are many reasons for this, and they range from technical (e.g., performance considerations) to economical ones. By no means do IN requirements agitate for moving all switching software outside of the switches.

- Gateway function GW1 for BCP-to-BCP internetwork access (when both networks are IN)

- Gateway function GW2 for Service Logic Process (SLP)-to-BCP internetwork access

- Gateway function GW3 for SLP-to-SLP internetwork access

- Gateway function GW4 for BCP-to-BCP internetwork access (when only one network is IN)

- Gateway function GW5 for the Service Management Process (SMP)-to-SMP internetwork access

A gateway relevant to service creation is absent in the preceding list. This is not an accident: it has been left unnamed intentionally. Recommendation Q.1201 states that the "need for network interworking at the service creation level is for further study." One may ask why these gateways have been mentioned at all, if they are not specified in detail. This is a valid question. The response to it is that by having agreed in principle that network interworking is a subject of IN standardization, the industry opened a door to the future contributions with specific technical proposals. Once such proposals are brought into standards, they will be judged on their technical merits, instead of being dismissed as irrelevant, which could have been the case had the recommendation omitted the description—as high-level and void of detail as it is—of network interworking arrangements. And the fact that the service creation gateway has been left unnamed is a nuance that signals to an experienced reader of standards documents that standardization of service creation per se is a taboo.[23]

Overall, none of the gateway functions has so far been specified in any detail, although some specific proposals have been made (Faynberg, 1995). As we will see later in this book, the (only) network interworking interface standardized in CS-1 needs no gateway at all. It is the authors' opinion that a gateway (whenever needed) should be specified in such a way that the existing (for the same function) intranetwork protocol can be reused.

As far as the IN LTA framework is concerned, Recommendation Q.1201 defines it as "the structure whereby there is integration of technologies developed in other standards activities." Such activities include the work of ITU-T Study Group 7, Question 19, "Distributed Architecture Framework (DAF)," which, in collaboration with the International Organization for Standardization (ISO) defines Open

---

[23]cf. Slavic languages: none of them has a word equivalent to "bear." The animal was feared so much that it could not be named; instead, it has been referred to, literally, as "the one who eats honey."

Distributed Processing (ODP). The ITU-T Study Group 4 project on Telecommunications Management Networks (TMN) is another key object-oriented modeling activity that will influence LTA. Yet another technology (or rather set of technologies) mentioned in Recommendation Q.1201 is *Computer-Aided Software Engineering* (CASE), which is expected to influence Service Creation. The LTA group within Group 11 meets regularly; so far no long-term Recommendations have been planned.

2.2.4 IN architectural concept

In proposing the IN Conceptual Model (INCM)—the subject of this section—Recommendation Q.1201 notes that it "should not be considered in itself an architecture..." but rather serve as "a framework for the design and description of the IN architecture."[24] This is an important note to keep in mind, for the authors have often heard questions like, "How do the four planes of the INCM relate to the OSI layers?" from the people who just started to learn IN. The answer to this question, of course, is that INCM does *not* relate to the OSI model. The latter is an architecture, but the INCM is just a set of *viewpoints* (using the ODP terminology). The INCM is depicted in Fig. 2.2 (which is a copy of Fig. 20 of Recommendation Q.1201) as a set of four "planes": Service Plane, Global Functional Plane, Distributed Functional Plane, and Physical Plane. These planes represent different aspects of implementing services as follows:

- The *Service Plane* deals with service specification (or, conversely, an observation of a service in action). Services are described in terms of *service features*. Recommendation Q.1201 gives no definition of this term (the discussion of service features is deferred until Recommendation Q.1202); for now, we assume that, as far as service specification is concerned, service features are verbal descriptions of what a service should (or should not) do in given circumstances. For example, a call queuing feature of an 800 Freephone[25] service may be described as follows: when all agents to whom the call may be directed are busy, the caller is put on hold until an agent is free, at which point the call is directed to this agent; if there is more than one caller waiting at this point, the first-come-first-served discipline is employed to select the caller to

---

[24]To this end, INCM has also served as the framework for the outline of this book.

[25]This term has been accepted internationally and used in ITU; earlier (Ambrosh et al., 1989), it had been also known as *Green Number Service* (GNS), but this evocative name has not been used much recently.

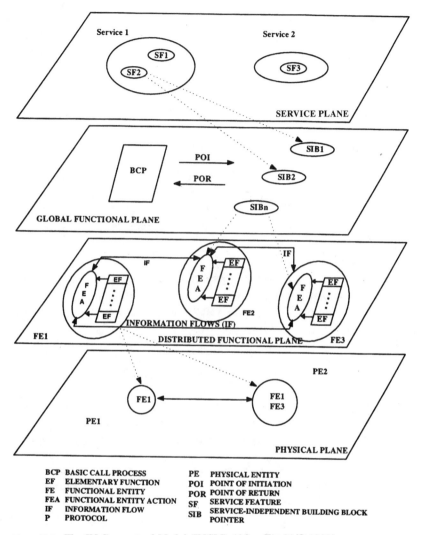

BCP BASIC CALL PROCESS
EF ELEMENTARY FUNCTION
FE FUNCTIONAL ENTITY
FEA FUNCTIONAL ENTITY ACTION
IF INFORMATION FLOW
P   PROTOCOL

PE PHYSICAL ENTITY
POI POINT OF INITIATION
POR POINT OF RETURN
SF SERVICE FEATURE
SIB SERVICE-INDEPENDENT BUILDING BLOCK
    POINTER

Figure 2.2   The IN Conceptual Model (INCM). (After Fig.20/Q.1201).

be connected to the available agent. Note that the principle of service independence also apply here, since service features that comprise the Service Plane may be used within the context of other services: for example, IN-provided call queuing features may be used equally well in the context of 900 (Premium Calling) service. Further, a service specification at that level does not take into account any aspects of the underlying network, so the Service Plane represents a service designer's viewpoint. Conversely, a service user typically "observes" the service at this plane.

- The *Global Functional Plane* is where the service is expressed in terms of Service-Independent Building Blocks (SIBs).[26] The SIBs are atomic instructions, which are chained together to form a Service Logic Program (SLP). The network *executes* SIBs in the following way: whenever the Basic Call Process (BCP)[27] (which is executed by the switching nodes of Fig. 2.1) passes the control to service logic [indicated by the arrow leaving the Point of Initiation (POI) in Fig. 2.2] the SIBs are executed in the way they are connected in the chain; in the end, control returns to the BCP, which, however, may continue its execution at a different point, the Point of Return (POR), rather than the POI. A service programmer, equipped with an ideal service creation package that shields him or her from the network, "observes" IN at this plane. But this plane's viewpoint may also be—quite fruitfully!—used as a step in service standardization as will be explained when we discuss the ITU-T methodology further in this section.

- The *Distributed Functional Plane* is composed of computational objects called Functional Entities (FEs). (None of these objects is tied with any piece of physical hardware, which is why the word "functional" is used.) The FEs may perform atomic Functional Entity Actions (FEAs), and, as the result of FEAs, exchange messages called Information Flows (IFs). An FE may send an IF only to those FEs to which it is connected[28] (this is a place in the model where the physical network connections are reflected) and even then only in a specified direction. At this point, more light can be shed on the SIB concept of the Global Functional Plane: SIBs are *realized* by a sequence of FEAs in specific FEs. Finally, Recommendation Q.1201 mentions Elementary Functions (EFs) within FEs, but they are left for further study and will never be mentioned again in this book. With respect to the viewpoint represented in this plane, it is that of a network designer.

- The *Physical Plane* is where the real network hardware resides. The Physical Entities (PEs) (i.e., switches, general-purpose computers that contain databases, etc.) of which the network is composed exchange *protocol* messages. The FEs of the Distributed Functional Plane are assigned to PEs, and the IFs between the communicating

---

[26]Recommendation Q.1201 does not actually define SIBs, which are discussed in Recommendation Q.1203.

[27]ITU-T Recommendations call BCP an SIB. It is merely a matter of definition, which the authors find confusing for several reasons (for example, BCP has none of the attributes that the other SIBs have), and therefore avoid.

[28]Recommendation Q.1201 actually does not mention this property of the model; this is done in Recommendation Q.1204.

FEs in *different* PEs are mapped into the protocol messages. Network and protocol designers "observe" IN at this plane.

Now we introduce the ITU-T (pre-IN) methodology for service design, and compare it with the INCM. In connection with ISDN studies in ITU-T, a three-stage methodology was developed to describe ISDN services and derive the protocols to support them. This process (which is still widely used today) is carried out, for each service, in three stages as follows:

1. Stage 1 describes the service as perceived by the user.

2. Stage 2 defines the capabilities and processes within the network that are required to provide the service. The output of this stage is the functional decomposition of the network components into FEs as well as the full specifications of the FEAs and IFs to support the service.

3. Stage 3 produces the protocol specification.

It is easy to see that these three stages correspond to, respectively, Service Plane, Distributed Functional Plane, and Physical Plane. In fact, IN has simply adopted these concepts, including the terminology associated with them (FEs, IFs, and so on). But the Global Functional Plane had no counterpart in the old terminology; it represents an entirely new and singularly important concept. Indeed, if the services can be programmed at the Global Functional Plane using a predefined set of SIBs and for each SIB there exists a standard protocol specification, there is no need to derive and maintain a separate protocol for each service! This is an illustration of the *mechanism* through which service independence (and network independence) is achieved. Whatever can be done at the Global Functional Plane is both service- and network-independent.[29]

2.3 An Interpretation of the Four-Plane IN Model in Light of Computer Architectural Concepts

The authors, who have participated in many discussions of the IN architectural concepts in general and INCM in particular, have

---

[29]As we will discuss later in this book, both CS-1 and CS-2 SIBs were standardized not for the purpose of service creation (which would have required specifics that many companies did not wish to standardize for competitive reasons) but that of service modeling. Presently, the Service Architecture Subworking Party in ITU-T is considering the acceptance of standardized IN SIBs, which may eliminate the need to redefine the individual protocol development for each service (Zeuch, 1996).

observed a common type of misunderstanding that has arisen because of a rather high level of abstraction and the seeming ambiguities of certain definitions. For example, the standard definition of an SIB is not (and probably cannot be) formal; thus, it may make a nonspecialist guess among different interpretations or abandon any attempt to understand the model altogether. This should not be the case.

To articulate this point and to build an intuitive understanding of key concepts such as SIB, we have developed a similar set of four viewpoints for the well-known von Neumann computer architecture. In conformance to the IN terminology, we call these viewpoints "Global Services Plane," "Global Functional Plane," "Distributed Functional Plane," and "Physical Plane." Then we explain how the IN architecture maps into these views. In other words, we describe the *computer domain* in terms already used to describe the *IN domain*. The model is depicted in Fig. 2.3.

There are two major benefits of this approach. First, it gives an immediate intuitive level of understanding of the basic concepts (the SIB concept being the most important of them), which is the next best thing to producing formal definitions. Secondly, it provides a direction for predicting possible future developments in telecommunications. It is an adage that similar problems have similar solutions.

Although there will be a few references to the material of this section in the rest of the book, most of it may be skipped without much harm. Its primary purpose is pedagogical: to explain the IN concept through computing concepts. To the extent the reader is familiar with the latter, he or she will find the material of this section useful, but the reader should feel no guilt in skipping it. On the other hand, if the reader finds the material useful, he or she may consult Faynberg et al. (1993) for a more detailed treatment of the subject.

A personal computer—even a portable one—is, in fact, a network of several units that communicate among themselves using complex multilayered protocols. Yet the users (and even the designers) of computer programs are rarely if at all aware of all aspects of this complexity.

For example, in order to write this section, the authors used several computer programs (e.g., editor and formatter), which implemented certain *services*. One program, an *editor,* provided editing services, which has such features as WYSIWYG (What You See Is What You Get) and the like. The other provided a printing service, whose execution resulted in a hard copy of the paper. And, of course, we used the same computer to invoke (by virtually pushing a button) other programs, which provided totally different services (e.g., spreadsheet and E-mail).

It is indeed possible today not only to use, but also to develop such utilities by being aware of only their features and some relatively high-level mechanisms of the features' invocation. (Again, using the ODP

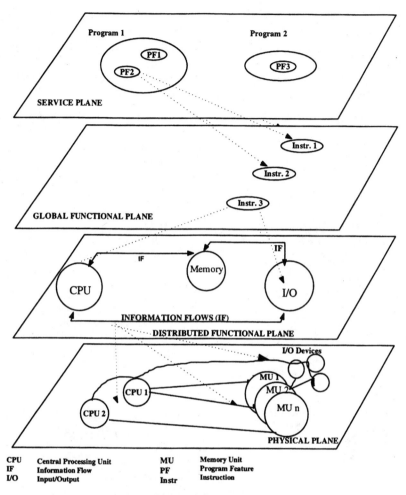

Figure 2.3 The computer conceptual model.

terminology, such a level of understanding of a computer represents a *viewpoint,* which we call "Service Plane.")

2.3.1 "Service Plane"

Most computers users are familiar only with this plane, which, in general, presents a view of the computer only in terms of specific programs that provide specific services; to receive these services one needs no knowledge of the computer's inner working. This happened by design; many researchers and developers worked to make computers easy to use. If more knowledge had been required, computers would have never penetrated our lives the way they have. Today, most computer users are familiar with (and depend on) almost universally portable programs. Again, to use any of these packages, one only has to learn the package's features.

So far, the discussion of the "Service Plane" has concentrated only on the users of computer applications. Of course, the designers and developers of these applications deal with the viewpoint provided at this plane, too. First of all, the writers of the requirements that describe the features of the system work in this plane. (In an ideal case, the requirements document does not differ from the user's manual.) Second, such requirements are usually straightforward to implement in the high-level languages that most application programmers use. In many cases, the tools and programming primitives hide from application programmers not only the structure of a particular computer but even that of any supporting software.

It rarely happens that application programmers have to write their programs in a machine language, but what a computer actually interprets when programs are executed is a set of *instructions* in such a language. Here comes the natural borderline between the issues within the viewpoint of what we define as the "Service Plane" and the next plane below, the "Global Functional Plane."

2.3.2 "Global Functional Plane"

At this plane the computer is still viewed as one unit. In addition, this viewpoint still concentrates on *what* is executed rather than on the *way* it is executed. But while the constituents of the "Service Plane" are an infinite combination of programs and their features, the "Global Functional Plane" is populated by a *finite* set of atomic building blocks, from which all programs are built. Such building blocks—the machine-level language instructions—share the following characteristics:

1. They are executed one at a time.
2. Each of them is uninterruptible.
3. Each of them specifies the following:

   - A generic operation (e.g., an arithmetic or logical operation or a transfer from one part of the memory to another) to be performed
   - Specific parameters for this operation as relevant to the program in whose context it is executed (e.g., numbers, strings, or memory addresses)
   - A pointer to the next instruction to be executed

A program can thus be viewed as a directed graph whose nodes are these instructions.

Our observation is that the IN term *Service-Independent Building Block (SIB)* denotes a parallel concept (i.e., a machine-level instruction, where the "machine" is the whole network), and so we will use it interchangeably with the term *machine-level instruction*. While it is true that the programmers who write at the machine level have to know quite a few things about the architecture of the computer they are working on, this knowledge rarely deals with more detail than the functional computer components (as explained in the next section) and never requires understanding of the protocol supported by these components to execute instructions.

The viewpoint considering the functional components of a computer and their interworking belongs to the "Distributed Functional Plane."

2.3.3 "Distributed Functional Plane"

The essential functional entities of a computer are as follows:

■ *Central Processing Unit (CPU)*. This is the heart of a computer. The CPU interprets the program instructions and executes them (cooperatively with other functional entities). The word *central* is used to indicate that most of the control functions are centralized in this entity.

■ *Memory*. This functional entity stores the data (including the program instructions themselves) and provides the access for data retrieval.

■ *Input/Output (I/O)*. Computers accept and dispense coded information by means of devices capable of handling such data. Actually, the name of this functional entity is a generic term for the set of devices (including disk, tape, modem, terminal monitor, terminal keyboard, and the like).

■ *Bus*. This is a fast communications network that connects the rest of the entities.

Most of the instructions, when executed by a CPU, require exchange of information with other functional entities. For example, a simple (from the machine-level programmer's point of view) instruction to move a byte of data from one location to another involves special protocols for acquiring the bus, exchanging handshake with the memory unit, etc.

There are two aspects to specification and implementation of these protocols. One aspect deals with the generic access protocol to a unit as an object. In other words, each unit has a specification for operations that can be performed on it as well as particular procedures that specify the sequencing for the proper operation of the unit. The other aspect deals with the use of these operations to implement a specific instruction. Given that the rules imposed by the rest of the units are obeyed, the flexibility in the development is entirely up to the designer of the CPU. There are surprisingly many ways to implement a single instruction; some are faster and more economical than others. But the instruction will always provide the same service, independent of its implementation.

Finally, we observe that with all the diversity in implementations of different functional entities in a computer (regarding the choices of memory and I/O devices), the interfaces between these functions are described in a way invariant to physical implementations. Thus the discipline of digital design (Mano, 1979) provides universally used practices for describing the interfaces to functional components (even the very basic ones) independent of the physical implementation of these components. The fact that computer components are pieces of electronic equipment is irrelevant as far as their functional description is concerned. Indeed, what used to be built with electromechanical relays, was later built with electronic tubes, and then transistors. Overall, it is the Distributed Functional Plane viewpoint that is critical as far as the overall design of computers is concerned. At what we call "Physical Plane," the functional entities of computers are implemented in specific pieces of equipment.

2.3.4 "Physical Plane"

A functional entity implemented 20 years ago in a piece of physical equipment the size of a hall can now be implemented in a pocket-size piece of silicon. As computers become smaller and more powerful, functional entities are grouped into single modules. Modern CPUs have their own internal memory, and memory management units and I/O devices are effectively full computers—they come on boards that contain CPU and memory. The networking inside of computers is also supported by several buses and, often, fast interconnection networks.

In other words, the mapping of functional entities into physical ones can be performed in many different ways. This diversity in today's multivendor environment exists thanks to the standards that define the interfaces between the functional entities. And, again, as long as two otherwise totally different computers maintain the same instruction (SIB) sets, all the software that runs on one machine will run on the other.

This completes our analogy between the computer domain and IN domain.

2.4   On the Role and Structure of Q.12x1 Recommendations

For each Capability Set x, the Recommendation Q.12x1 by design constitutes a lawyer's agreement of sorts on what is (and what is not) standardized in this particular Capability Set. One problem these recommendations pose for the authors is that it is hard to locate a single place in the book to discuss them. Even though it is the role of the Q.12x1 recommendations to select and elucidate the principles and requirements specific to CS-x, for which reason the discussion of these recommendations belongs in the present chapter, the material of these recommendations builds on that of the rest of general IN recommendations (Q.1202, Q.1203, Q.1204, Q.1205, and Q.1208) and thus cannot be included in the present chapter without breaking the natural flow of the book. In other words, the discussion of, for example, CS-1 specifics in this chapter is impossible without making forward references. For this reason, we took another approach in discussing Q.12x1 recommendations.

As the rest of the book reviews the standards top-down, IN plane by IN plane, it appears natural to split the discussion of Q.12x1 recommendations into four parts, with each part relevant to a specific plane. Only Q.12x1 recommendations are treated this way, but then it is only Q.12x1 recommendations that span all four planes of IN!

Chapter

3

IN Service Plane

3.1 Overview

It often happens in present technical publications that references to
"IN services" are made. This term usually denotes services that are traditionally implemented using IN (such as Freephone or VPN), but the
term is misleading. There is simply no such thing as an "IN service,"
because IN is service-independent: any service, in principle, can be
implemented using IN or without using IN. IN does *not* standardize or
define services but does define and standardize the means of supporting their introduction into networks. Therefore, a much better term to
use when speaking of services in connection with IN is IN-based or IN-
supported.

What is the role of the IN Service Plane then? Mainly it is to keep a
repository of the services that can be supported by a given IN
Capability Set. As Chap. 1 indicates, by the time IN standardization
started, there had been several existing IN implementations (and several proposals for the new ones) that had significant differences in their
capabilities and underlying protocols. An attempt to amalgamate these
protocols into one, working bottom-up, did not seem to foster agreements, for which reason the standardization process was reorganized
to include the top-down approach. The latter—and it has proven to be
successful—first builds an agreement on the services and service features to be supported in a given Capability Set and only then considers
the lower-level capabilities needed for the support of the services chosen. Thus it is perfectly correct to speak about CS-1 services, CS-2 services (which, of course, include CS-1 services), and so forth.

As far as Capability Sets are concerned, in CS-1 the service principles are reflected only in Recommendation Q.1211, but in CS-2 studies

it was recognized that the amount of service material[30] warranted a separate recommendation to cover its service aspects. It is very likely that this trend will continue in the future Capability Sets. This chapter is organized to reflect the purpose of the Service Plane as the repository of services and service features relevant to specific Capability Sets, as follows:

- Section 2 is dedicated to Recommendation Q.1202.

- Section 3 describes the service-specific aspect of Recommendation Q.1211.

- Section 4 shifts to CS-2 by reviewing the service-specific aspects of draft Recommendation Q.1221.

- Section 5 discusses Recommendation Q.1222.

3.2 Recommendation Q.1202, Intelligent Network—Service Plane Architecture

3.2.1 Summary

Recommendation Q.1202 is only three pages long. The first (and still current) publication of the recommendation is dated October 1992. This recommendation was developed jointly with ITU-T Study Group 18, which assigned it its second name, I.328.

Recommendation Q.1202 defines the concepts of service and service features as well as the concept of service interaction. Finally, the recommendation specifies the methodology for identifying SIBs.

3.2.2 Services and service features

The concepts of service and service features are respectively defined as follows:

- "A *service* is a stand-alone commercial offering, characterized by one or more core service features, and [it] can be optionally enhanced by other service features"

- "A *service feature* is a specific aspect of a service that can also be used in conjunction with other services and service features as part of [a] commercial offering. It is either a core part of a service or an optional part offered as an enhancement to a service."

---

[30]In the beginning of the ITU-T study period, the material destined for standardization accumulates in so-called "living documents" or "baseline documents." Depending on the collected material, relevant Recommendations are named and drafted.

Recommendation Q.1202 implicitly makes an important point (already mentioned in the overview of this chapter): there are no IN services but there are IN-*supported* services. To this end, Q.1202 postulates that "the [Service Plane] view contains no information whatsoever regarding the implementation of the services in the network (for instance, an IN type of implementation is invisible)."

As general as the preceding definitions seem to be, it would be hard to produce more specific definitions without resorting to formal methods (based, for example, on mathematical logic). The definitions do capture, however, the commercial meaning and interrelation of the terms "service" and "service feature" and indicate that the terms relate to the perception of the end user rather than any implementation mechanism.[31]

3.2.3 Service interaction

Because IN handles multiple instances of services that can be activated concurrently for the same call, certain service features that are perfectly acceptable within a given service may be contradictory to the features of other services. For example, a user can subscribe to a feature called *abbreviated dialing,* which will enable this user to dial a number by entering a wildcard character (say, "*") followed by a two-digit index into the user's directory of full numbers. If a service that requires screening the dialed numbers is invoked, the full number may be unavailable, in which case the user may be denied the service. In general, when services behave in a way undesirable to the users (and unforeseen by the designers), one observes *service interactions.*[32] We will address this phenomenon later in this chapter; for now, we concentrate only on those interactions that are known to service designers beforehand. Even in this case, eliminating their occurrence is a nontrivial problem. Recommendation Q.1201 notes the agreement that in an "IN-structured network, service interactions can be customized" and gives a useful example of such customization to eliminate the occurrence of service interactions for Freephone and Call Forwarding Unconditional services.

---

[31]In this connection it is interesting to observe the development of wireless networks in the U.S., where each network provided users with certain features. As the users crossed the network boundaries, they seemed to have lost the features they were accustomed to in their home networks. Keeping the users' perception of seamless service meant providing identical features across the networks—the driving reason for the development of Wireless Intelligent Network (WIN) architecture.

[32]Often the term *feature interactions* is used in technical literature, which refers to the same phenomenon at a different level of granularity.

3.2.4   Identification of Service-Independent Building Blocks (SIBs)

Recommendation Q.1202 proposes the iterative methodology for identification of SIBs, which is the first necessary phase in ensuring that the IN infrastructure indeed supports a given set of services. The methodology is composed of the following three steps:

1. Identify a set of services and service features.

2. Select or modify a set of existing SIBs.

3. Verify that all services work with this set of SIBs. If not, restart with step 2.

Note that the methodology spans two IN planes: the Service Plane and Global Functional Plane.

3.3   Service Aspects of Recommendation Q.1211, Introduction to Intelligent Network Capability Set 1

Recommendation Q.1211 states that CS-1 supports service features (and, consequently, services) that are single-ended and have a single point of control. Those criteria are defined in the recommendation as follows:

1. "A *single-ended service feature* applies to one party in a call and is orthogonal (independent) at both the service and topology levels to any other parties that may be participating in the call. Orthogonality allows another instance of the same or a different single-ended service feature to apply to another party in the same call as long as the service feature instances do not have feature interaction problems with each other."

2. "*Single point of control* describes [a] relationship where the same aspects of a call are influenced by one and only one Service Control Function at any point in time."

In other words, *single-endedness* means that a service process deals with only one call party[33] and having a *single point of control* means that service control processes on different machines don't have to know anything about each other, including each other's existence. (The con-

---

[33]This term refers to both the calling party (i.e., the party that initiates a call) and the called party.

cept of *Service Control Function* belongs in the Distributed Functional Plane; for the purposes of this chapter, it refers to the computing process that controls the service.)

Recommendation Q.1211 calls the services that satisfy the above criteria Type A services; all the rest are called Type B services.

The target set of CS-1-supported services, as defined in Recommendation Q.1211, is as follows:

Abbreviated Dialing	(ABD)
Account Card Calling	(ACC)
Call Distribution	(CD)
Call Forwarding	(CF)
Call Rerouting Distribution	(CRD)
Credit Card Calling	(CCC)
Destination Call Routing	(DCR)
Follow-Me Diversion	(FMD)
Freephone	(FPH)
Malicious Call Identification	(MCI)
Mass Calling	(MAS)
Originating Call Screening	(OCS)
Premium Rate	(PRM)
Security Screening	(SEC)
Selective Call Forward on Busy/Don't Answer	(SCF)
Split Charging	(SPL)
Televoting	(VOT)
Terminating Call Screening	(TCS)
Universal Access Number	(UAN)
Universal Personal Telecommunications	(UPT)
User-Defined Routing	(UDR)
Virtual Private Network	(VPN)

Two more services, Completion of Call to Busy Subscriber (CCBS) and Conference Calling (CC), are identified with the note that they actually require additional Type B capabilities, and therefore can be only partially supported in CS-1.

Annex B of Recommendation Q.1211 provides the prose description of targeted services and service features. (In most cases, there are several such descriptions, possibly inconsistent with each other, as the disclaimer at the beginning of the document states.)

In addition to the above set of services, the set of target service features has been also identified:

Abbreviated Dialing	(ABD)
Attendant	(ATT)
Authentication	(AUTC)
Authorization Code	(AUTZ)
Automatic Call Back	(ACB)
Call Distribution	(CD)
Call Forwarding	(CF)
Call Forwarding on Busy/Don't Answer	(CFC)
Call Gapping	(GAP)
Call Hold with Announcement	(CHA)
Call Limiter	(LIM)
Call Logging	(LOG)
Call Queuing	(QUE)
Call Transfer	(TRA)
Call Waiting	(CW)
Closed User Group	(CUG)
Consultation Calling	(COC)
Customer Profile Management	(CPM)
Customized Recorded Announcement	(CRA)
Customized Ringing	(CRG)
Destination User Prompter	(DUP)
Follow-Me Diversion	(FMD)
Mass Calling	(MAS)
Meet-Me Conference	(MMC)
Multi-Way Calling	(MWC)
Off-Net Access	(OFA)
Off-Net Calling	(ONC)
One Number	(ONE)
Origin-Dependent Routing	(ODR)
Originating Call Screening	(OCS)
Originating User Prompter	(OUP)
Personal Numbering	(PN)
Premium Charging	(PRMC)
Private Numbering Plan	(PNP)
Reverse Charging	(REVC)
Split Charging	(SPLC)
Terminating Call Screening	(TCS)
Time-Dependent Routing	(TDR)

Of these service features, ACB, CHA, TRA, CW, COC, MMC, and MWC may not be fully supported in CS-1 because they require Type B capabilities. Annex A of Recommendation Q.1211 contains informative tables, which identify the service features for each service.

3.4 Service Aspects of (Draft) Recommendation Q.1221, Introduction to Intelligent Network Capability Set 2

The present text of this recommendation (Lu, 1996) provides the set of benchmark services and service features "to be used to identify and verify the service-independent capabilities of CS-2." Certain services are declared "essential," while the others are considered optional by the recommendation.[34] The essential features are already supported in the refined CS-1. [The *Universal Personal Telecommunications* (UPT) Set 1 features, for example, are essential.]

As far as the services not supported in CS-1 are concerned, Recommendation Q.1221 introduces[35] "the limited functionality of mobility services and B-ISDN services."

The recommendation also addresses the capabilities supporting implementation of service management and service creation.

The telecommunications services to be supported in CS-2 (in addition to all CS-1-supported services, naturally) include

Internetwork Freephone	(IFPH)
Call Transfer	(TRA)
Internetwork Premium Rate	(IPRM)
Call Waiting	(CW)
Internetwork Mass Calling	(IMAS)
Hot Line	(HOT)
Internetwork Televoting	(IVOT)
Multimedia	(MMD)
Global Virtual Network Service	(GVNS)
Terminating Key Code Screening	(TKCS)
Completion of Call to Busy Subscriber	(CCBS)
Message Store and Forward	(MSF)
Conference Calling	(CONF)

---

[34]The example description of the services and service features can be found in Annex B of Recommendation Q.1221; the mapping of services to service features is contained in Annex A.

[35]In clause 5.

International Telecommunication Charge Card	(ITCC)
Call Hold	(CH)
Mobility services	
Universal Personal Telecommunications	(UPT)
Future Personal Land Mobility Telecommunications Services	(FPLMTS)

The supported (or partially supported) service features are

User Authentication	(UAUT)
Inter-Network Service Identification	(INSI)
User Registration (UREG)/Outgoing Call Registration	
Inter-Network Rate Indicator, Forward	(INRI-F)
Secure Answering	(SANSW)
Inter-Network Rate Indicator, Backward	(INRI-B)
Follow-on	(FO)
Real Time Flexible Rating	(RTFR)
Flexible (Call) Origination Authorization	(FOA)
Originating Carrier Identification	(OCI)
Flexible (Call) Termination Authorization	(FTA)
Terminating Carrier Identification	(OTC)
Provision of Stored Messages	(PSM)
Resource Allocation	(RAL)
Multiple Terminal Address Registration	(MTAR)
Delivery of Complementary Information	(DCI)
Intended Recipient Identity Presentation	(IRIP)
Service Indication	(SIND)
Blocking/Unblocking of Incoming Calls	(BUIC)
Service Negotiation	(SNEG)
Terminal Authentication	(TAUT)
Call Forwarding	(CF)
Handover	
B-ISDN Multiple Connections Point to Point	(BI-MCPP)
Terminal Location Registration	(TLR)
B-ISDN Multi-Casting	(BI-MCAST)
Terminal Attach/Detach	(ATDT)
B-ISDN Conferencing	(BI-CONF)
Terminal Paging	(TPAG)
Call Connection Elapsed Time Limitation	(CCEL)

Radio paging	(RPAG)
Special Facility Selection	(SFS)
Emergency Calls in Wireless	(ECW)
Concurrent Features Activation with Bi-Control	(CFA-BC)
Terminal Equipment Validation	(TEV)
Customised Call Routing with Public network	(CCR-PU)
Cryptographic Information Management	(CIM)
Customized Call Routing with Private network	(CCR-PR)
Automatic Call Back	(ACB)
Internetwork Service Profile Interrogation	(ISPI)
Call Hold	(CH)
Internetwork Service Profile Modification	(ISPM)
Call Retrieve	(CRET)
Internetwork Service Profile Transfer	(ISPT)
Call Transfer	(CTRA)
Reset of UPT Registration for Incoming Calls	(IRUR)
Call Toggle	(CTOG)
Mobility Call Origination (Mobile Call Origination/UPT outcall)	(MCO)
Call Waiting	(CW)
Mobility Incall Delivery (Mobile User Call Term./UPT incall delivery)	(MID)
Meet-Me Conference	(MMC)
Data Communication between Different Protocol Terminals	(DCPT)
Multi-Way Calling	(MWC)
Charge Determination	(CDET)
Call Pick-Up	(CPU)
Charge Card Validation	(CCV)
Calling Name Delivery	(CND)
Call Disposition	(CD)
Services On-Demand	(SOD)
Enhanced Charge Card Validation	(ECCV)
Message Waiting Indication	(MWI)
Enhanced Call Disposition	(ECD)
Feature Use Charging	(FUC)
User Service Interaction	(USI)

Recommendation Q.1221 introduces the service management and service creation capabilities (both of which are also referred to as ser-

vices) for the first time in the Q.12*xx* series. The introduction starts with the definition of the roles of the personnel accessing these capabilities. To this end, the roles of service subscribers and service administrators are defined[36] as follows:

> "*Service subscribers* are users of the IN-supported services. (When subscribing for a particular IN-supported service, the service subscribers may choose some service features offered by the service provider according to their needs.) *Service administrators*...[represent] both the service providers and network providers; they are responsible for the following areas of service support: service provisioning..., service maintenance..., operations management..., systems security..., and network administration...."

The recommendation further notes that the "above activities so far have been performed only by service administrators; however, service subscribers need and demand more control over the specifics of the services." Responding to this demand, the recommendation distinguishes the capabilities of the service subscribers to control the services they are subscribed to, which it groups under the name "Customer Control Service Capability." A set of such capabilities to be given to a particular user is service-dependent, but, in general, these capabilities allow the service subscribers to view and change certain service parameters and receive appropriate reports.

The Service Management Services include three major groups, Service Customization Services, Service Monitoring Services, and Service Control Services, which are as follows:

1. *Service Customization Services.*   Telecommunications Service Customization (TSC), Service Control Customization (SCC), and Service Monitoring Customization (SMC)

2. *Service Monitoring Services.*   Subscriber Service Report (SSR), Billing Report (BR), Subscriber Service Status Report (SSSR), Subscriber Traffic Monitoring (STM), and Subscriber Service Management Usage Report (SMPUR)

3. *Service Control Services.*   Subscriber Service Activation/ Deactivation (SSAD), Subscriber Monitoring Activation/ Deactivation (SMAD), Subscriber Profile Management (SPM), Subscriber Service Limiter (SSL), and Subscriber Service Invocation (SSI)

The rest (i.e., the ones that do not belong to any of the above three categories) of the service management services are Subscriber Service

---

[36]In clause 5.2.

Testing (SST), SMP Usage Report (SUR), and Subscriber Security Control (SSC).

Finally, the service creation services are grouped into the five categories, *Service Specification Services, Service Development Services, Service Verification Services, Service Deployment Services, and Service Creation Management Services,* as follows:

1. *Service Specification Services.* Feature Interaction Detection (FID),[37] Cross-Service Feature Interaction Detection (CSFID), Feature Interaction Rule/Guidelines Generation (FIRGG), Service and SIB Cataloguing (SASC), and Created Service Resource Utilization (CSRU)

2. *Service Development Services.* Creation Interface Selection (CIS), Creation Initiation (CI), Editing (ED), Combining (CO), Data Population Rule Generation (DPRG), SMP Service Creation (SSC), Syntax and Data Checking (SDC), Service and SIB Archiving (SASA), Service Configuration Control (SCC), SIB Configuration Control (SIBCC), and Network Configuration Tracking Capability (NCTC)

3. *Service Verification Services.* SCE Testing (ST), Created Service Simulation (CSS), and Created Service Live Testing (CSLT)

4. *Service Deployment Services.* SMP-Created Service Data and SLP Update (SCCD/SU), Service Distribution (SD), SIB Distribution (SIBD), Data Rule Distribution (DRD), Feature Interaction Rule Distribution (FIRD), Multiple SMP Support (MSS), Network Tailoring (NT), Network Element Capability Specification (NECS), and Network Element Function/Capability Assignment (NEF/CA)

5. *Service Creation Management Services.* SCE Access Control (SAC), SCE Usage Scope (SUS), SCE Recovery (SR), SCE Release Management (SRM), SCE Capability Expansion (SCECE), SCE Conversion (SCEC), Cross-SCE Service Maintenance (CSSM), SCE-to-SCE System Consistency (SCESC), SCE Service/Modular/System Transference (S/M/ST), Conversion of Created Services (COCS), and Service Management Interaction (SMI)

3.5 Draft Recommendation Q.1222, Service Plane for Intelligent Network Capability Set 2

In its most recent version (Hilton, 1996), the recommendation is five pages long. Its objective is to "provide the architecture of the IN CS-2

---

[37]The treatment (as limited as it is) of the feature interactions is provided in Recommendation Q.1222, which is discussed in the next section of the present chapter.

Service Plane...[so] that specific functionalities and their interactions can be identified and described in other Recommendations...."

The Recommendation provides definitions of the following concepts: feature interaction, feature cooperation, and feature interference. *Feature interaction* is defined as a "situation that occurs when an action of one feature affects an action or capability of another." It is noted that such a situation may be undesirable (i.e., when it disrupts a service) or desirable (i.e., when features "cooperate" in achieving the expected result). A desirable feature interaction is called *feature cooperation*; an undesirable one, *feature interference.*

In general, the interactions may occur not only among the *call-related* features (i.e., the features of the services or parts of the services that are performed within a given call context) but also among the *non-call-related* features (which are particular to CS-2—especially, its mobility aspects—and include user authentication and registration). The recommendation prescribes that the features of either type (as well as across both types) be examined on the subject of interactions.

The recommendation identifies two mechanisms that can be employed in handling feature interactions: the *static* and *dynamic* mechanisms. The former mechanisms are to be used in the design phase; the latter ones are to be used at run time. The specifics of the mechanisms are not presented in the recommendation.

While CS-1 addressed only Telecommunications Services, CS-2 also addresses Management Services (which include service customization, service control, and service monitoring) and Service Creation Services (which include service specification, service development, service verification, and service deployment).

Recommendation Q.1222 upholds the methodology for service description based on the INCM model and explains its relations to the *Unified Functional Methodology* (UFM) (Zeuch, 1996): the stage 1 service description is achieved through decomposition of services into service features, which are, in turn, expressed through SIBs.

Chapter

4

IN Global Functional Plane

4.1 Overview

The Global Functional Plane (GFP) provides the view of service developers. Recalling the computing metaphor explored in Chap. 2, we repeat that application programmers should not be concerned with the precise structure of the computers for which their programs are written. Even though the execution of programs always results in rather complex communications among the parts of a computer on which they are run, programmers, in general, do not have to be aware of these communications. Furthermore, a program written in a specific instruction set can run on architecturally different machines as long as their processors support this instruction set.

Similarly, a service composed of the *instructions* defined within the IN GFP[38] can be executed in all IN-structured networks that support such instructions. In other words, the GFP models the IN programming environment. We use the word *models* because the IN standards intentionally avoided specification of such an environment. [Doing so would have involved much more detail than the CS-1 schedule could permit, and, after all, the main goal of CS-1 was to standardize the protocol. In addition, as became clear during the CS-2 studies, some vendors felt that the IN programming environment (or, to use the IN terminology, the *service creation environment*) was a matter of competition rather than standardization.]

---

[38]As we mentioned before, these instructions are called *Service-Independent Building Blocks* (SIBs) in IN standards.

Nevertheless, SIBs do form an essential part of most service creation environments advertised today. Many recent publications on this subject by both service providers and vendors[39]

1. Report on implementation of CS-1 SIBs in commercial or research service creation environments

2. Describe the limitations of CS-1 SIBs

3. Offer suggestions for enlarging the SIB palette[40]

As the publications demonstrate, most vendors implemented all CS-1 SIBs, but they claim that many more are needed. To this end, CS-2 has taken steps to define new SIBs as well as to improve the existing ones. In addition, CS-2 is adding new constructs that allow creation of new (high-level) SIBs from the existing ones and support the concurrent execution of SIBs.

Again, in both CS-1 and CS-2, the primary role of SIBs has been to aid service modeling and service description. We discussed this subject in Chap. 2 in sufficient length, but it may still be worth pointing out that service designers would spend much less time if they could specify services without deriving service-dependent protocols. Instead, they could use SIBs; since the protocol to support each SIB is defined and standardized, the protocol to support the whole service can be derived automatically (e.g., by a compiler).

Recommendation Q.1219 contains superb self-explanatory examples of using SIBs for service description [for such services as *Universal Personal Telecommunications* (UPT) Set 1]; these service examples are followed through all the way to the protocol description.

The rest of this chapter reviews the ITU-T Recommendations Q.1203, Q.1213, and Q.1223 (draft), in that order. There are no GFP-specific requirements in either Recommendation Q.1211 or Q.1221.

4.2   Recommendation Q.1203, Intelligent Network—Global Functional Plane

4.2.1   Summary

Recommendation Q.1203 is 12 pages long. The first (and still current) publication of the Recommendation is dated October 1992. This

---

[39]Here is a list of only a few, most recent, publications: Ai et al. (1995), Akihara et al. (1995), Chang (1994), Clarisse et al. (1994), Elmgren and Majeed (1994), Gulzar and Salm (1995), Ku (1994), Marks et al. (1995), Petruk (1994), Squires (1995), Spencer (1995), and Syrett et al. (1995).

[40]In most implementations, SIBs are represented by their graphical interface icons. These icons form what is often called a SIB *palette.*

Recommendation is the last in the trilogy[41] developed jointly with ITU-T Study Group 18, which assigned it its second name, I.329.

The concepts considered in this Recommendation are those of the BCP, GSL, and SIBs. Each of these is discussed in a separate section following.

4.2.2   Basic Call Process

Basic Call Process (BCP) represents a non-IN (i.e., switch-based) part of activities that support a single call. It is important to note that the word *process* is used here to denote a distributed (from a computational point of view) activity. Consider a simple telephone call between two parties connected to the same switching exchange. When the calling party initiates the call, the exchange starts the *originating* process. When the exchange has enough information to ring the called party, it also starts the *terminating* process. In this case, the BCP actually combines these two processes. Of course, as far as the GFP is concerned, the distributed nature of the BCP is irrelevant; however, it is essential for a reader to be aware of it to fully understand the concept and appreciate its complexity.

The analogy with the computer domain can be quickly drawn as follows. Consider a human user. The user types a request at a terminal, which results in starting a process in the computer to which the terminal is attached. This process (which corresponds to the BCP in our analogy) invokes and executes a program on behalf of the user.

The following discussion is accompanied by Fig. 4.1, which is a slightly modified version of Fig. 5 of Recommendation Q.1203. Whenever the BCP is unable to provide a requested service by itself, it initiates (invokes) a program called *Global Service Logic* (GSL), whose execution constitutes another process. This happens at a point in BCP that is called the *Point of Initiation* (POI). Even though the call processing in the switching exchanges does not remain idle when the foreign logic is invoked, for the purposes of modeling this activity (and relevant software design), the execution of the BCP is considered suspended at the POI. When the execution of GSL is completed, the execution of the BCP is resumed.

Here is a potentially confusing issue. Recommendation Q.1203 effectively states that the BCP is resumed (after the execution of GSL terminates) at another point, which is called *Point of Return* (POR). All the figures in standards documents that depict the BCP and its relation to GSL (cf. Fig. 4.1) demonstrate that a POR may be different from a POI.

---

[41]The trilogy also includes Recommendations Q.1201/I.312 and Q.1202/I.328.

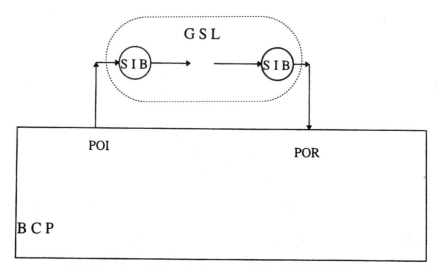

Figure 4.1   Global Functional Plane Architecture. (*After Fig. 5 / Q.1203.*)

Although most engineers—especially the ones who already know the requirements the model has to meet—intuitively understand what this means and accept the model, the computer specialists (especially the ones responsible for implementing the model in some form of software) raise a question at this point: how can a process be suspended at one point and be resumed at another one? Indeed, in a textbook model of process execution (Tanenbaum, 1992; Peterson and Silberschatz, 1985), a process is *always* be resumed at the point it was suspended, whether it had been suspended at its own demand (which is the BCP case) or because of some external activity asynchronous with the process execution. The BCP type of suspension happens[42] through some type of a procedural call [more precisely, a remote procedure call, which is well described in Tanenbaum (1989)]. But with the procedure call, the returning point is just the next instruction following the procedure call itself; in other words, the POI *must* be equal to POR!

To remedy the problem, we first note that in IN the points (or states) in BCP that may become POIs and PORs are well defined and, in fact, standardized.[43] For some service features and general service options (e.g., error handling cases), it is indeed necessary that the service logic programs be able to request a jump to another state in call processing, actually making the BCP start a new life rather than resume the old

---

[42]This detail is made clear in Recommendation Q.1208; however, it is needed to fully understand the model of Recommendation Q.1203.

[43]The underlying state machines are discussed in detail in the next chapter.

one. This is a valid requirement, and there are several techniques of modeling (and implementing) the software to meet it.

A good example of a similar model is software debugging tools that allow programmers to suspend their programs at some point so that they can inspect or modify the values of certain variables and then restart the programs at any point. The requirement to provide for an unusual manipulation of a process is met here, but, in this case, there is a special entity—practically, a part of the operating system—that performs such a manipulation. A similar entity, if added to the model, would make it much more clear.

To follow the analogy with operating systems,[44] we call this new entity the *monitor.* Figure 4.2 depicts the modified GFP architecture.

To demonstrate how this model works, we present one realization of it (a reader may find different—but still valid—realizations). Here the monitor is the controlling process, while the BCP is a *subroutine* with two formal parameters that respectively correspond to the state of the BCP and the event that has to be processed. Initially, the monitor calls the BCP with its starting state. The BCP returns to the monitor at the POI (determined by the event). The monitor then initiates the GSL, also via a (possibly remote) procedure call. When the GSL returns, its output is captured by the monitor, which translates it into the appropriate state (POR) of the BCP and the event to be processed, and then calls the BCP subroutine again, with the POR and the processed event passed as parameters. Please note that the model described here is *not*

---

[44]Whose early versions were implemented through what were called *monolithic monitors.*

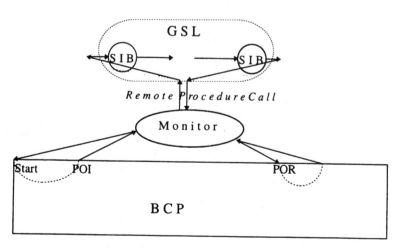

Figure 4.2   GFP interpretation model.

part of a standard, but rather our interpretation of it, which, we hope, will aid service developers.

One final note: in ITU-T Recommendations the BCP is referred to as a SIB. Perhaps it makes sense to call the BCP a SIB in that it represents a service-independent computational item; however, as we will see later in this chapter, the BCP is so much different from the other SIBs in both its form and its use that the word *SIB* appears to be a misnomer in this case.

4.2.3 Global Service Logic

The Global Service Logic (GSL) is defined in Recommendation Q.1203 as the "glue" that specifies the order of its instructions[45] (SIBs) as well as BCP's PORs to return to, and is responsible for holding the global data that relates to the call instance to be processed. This is about as much as can be said about any program, and GSL is just that—a program. The only difference is that GSL is cooperatively executed by different pieces of IN equipment; however, neither GSL nor its programmers are aware of it. As often happens with the names of programs, the word *GSL* is often used to denote one of the following:

1. *Programs*—symbols or statements written in a programming language by programmers

2. *Executable code*—statements written either in a machine language or the code to be recognized by an interpreter (in the case of GSL, the executable code consists of several modules, destined for different machines)

3. *Processes* (i.e., programs in execution)

What exactly is meant by the word *GSL* can, in most cases, be easily ascertained from the context, which may change, however, even within one sentence. For example, the first sentence of this section has two clauses; the first clause clearly speaks about GSL as a program, while the second one refers to its execution properties, thus discussing a process.

This subtlety should never be ignored when interpreting the standards documents. Consider the following. Some figures in Recommendations Q.1203 and Q.1219 depict more than one POR for a given POI. At an early stage of reviewing these documents that fact caused much confusion. We encountered many developers who thought that multiple PORs implied some complex execution scheme that somehow

---

[45]Here is one argument against calling the BCP a SIB: the BCP (or its symbol) is not part of GSL.

eluded their understanding,[46] and so they kept requesting more information from the authors. This situation, however, ought to be interpreted in only one way: an invocation of the GSL results in returning to one and only one POR. This POR, however, may depend on what happens at run time, for which reason the program must take into account *all* possible outcomes, each of which may result in different PORs.[47] Hence, multiple PORs are *programmer's* alternatives that are essential in the context of GSL as a program, but cease to exist in the other context.

4.2.4 Service-Independent Building Blocks (SIBs)

Recommendation Q.1203 (subclause 3.1) provides the following definition:

> "A SIB is a standard reusable network-wide capability residing in the Global Functional Plane used to create service features. SIBs are of a global nature and their detailed realization is not considered at this level but can be found in the Distributed Functional Plane (DFP) and the Physical Plane. The SIBs are reusable and can be chained together in various combinations to realize services and S[ervice] F[eature]s in the Service Plane. SIBs are defined to be independent of the specific service and technology for which or on which they will be realized."

It may be very hard to improve this definition; however, it may appear to an uninitiated user too abstract to pin down. To help in doing so, we go back to the analogy with the computer architecture that we first explored in Chap. 2. Consider an assembly-level programmer of a computer, who has at his or her disposal a set of assembly-level instructions (such as the ADD and COMPARE instructions of Table 4.1).

The former instruction adds two numbers (which we assume, for simplicity, to be always contained in the parameters) and stores the sum in the first parameter of the instruction. The latter instruction compares two numbers (contained in its first two operands) and then, depending on the outcome of the comparison, jumps to one of the addresses indi-

---

[46]Indeed, a procedure returns only once to its caller.

[47]Similarly, a program written in the C language may use several *return* statements, but one and only one of them will have an effect during the execution of the program.

TABLE 4.1 Two Assembly-Level Instructions

ADD	OPERAND 1	OPERAND 2			
COMPARE	OPERAND 1	OPERAND 2	EQUAL	GREATER	LESS

cated in the respective operand [i.e., EQUAL, LESS (than), or GREATER (than)] of the three remaining ones.

Using the SIB's definition terminology, these instructions can be called "computer-wide" because they are executed by the whole computer, as far as the programmer is concerned (i.e., the programmer does not have to know the intricacies of the interconnection of the modules with which the computer is built or the messages they exchange to execute each instruction.) These instructions are also the only means of implementation of program features available to the programmer. These instructions are reusable: each of them can appear more than once in a given program (and executed more than once within a process).

This is precisely what SIBs are to service developers: they are network-wide, reusable, and they are used to create service features.

Traditionally, assembly-level programs are written as a sequence of instructions, one instruction after another; however, the sequence in which they are written does not necessarily define the order in which they are executed. Consider the program in Fig. 4.3, which increments the value of variable i (assumed to be initially equal to zero) exactly 11 times.

The COMPARE instruction has three pointers to (alternative) instructions to be executed next, depending on the outcome of the comparison. In the example program, if the comparison reveals that i is less than 11, the COMPARE instruction requests (by means of specifying a label called INCREMENT as the value of its LESS parameter) that the ADD in front of it (which is labeled "INCREMENT") be executed; otherwise, the label CONTINUE refers to another instruction to be executed next. The ADD instruction specifies no label, which means—as far as most assembly languages are concerned—that the next instruction to be executed is simply the next instruction in the program text (the COMPARE instruction, in this case). Overall, there is a clear (although possibly run-time-dependent) order in which the instructions are "chained." At the time assembly languages were first

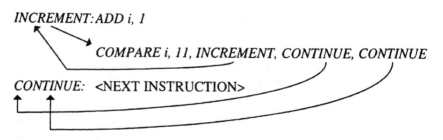

Figure 4.3 Example assembly-level program.

developed, they very closely reflected the machine languages,[48] in which the labels correspond to specific memory addresses of the instructions. Had graphical user interfaces been available at that time, the labels (and references to them) would most likely have never existed as a programming construct; instead, the "chaining" would have been performed graphically, for example, via arrows, as depicted in Fig. 4.3. With this, a label (or address) of an instruction would be represented by an arrow pointing at the instruction icon, and the branching choices for the instructions that could be executed next[49] would be represented by the arrows exiting the instruction icon. An implementation of this paradigm results in presenting the service logic in a form of a graph, often called the *decision graph* (Morgan et al., 1991).

Recommendation Q.1203 prescribes exactly this type of graphical depiction of SIBs (presented in Fig. 4.4, which is a copy of Fig. 3 of Recommendation Q.1203). Each SIB has a *Logical Start* and one or more *Logical Ends*,[50] each of which is connected to the next SIB in the chain. It should be clear now how this depiction is to be interpreted and what the word "chaining" in the definition of SIBs means.

What is left to be discussed is how SIBs exchange information with each other and the outside world. Again, similar to assembly-level

---

[48]Neither of our two example instructions, ADD or COMPARE, belongs to a particular assembly language or reflects any particular machine language.

[49]Again, there may be only *one* next instruction as far as the execution is concerned.

[50]Sometimes (e.g., in figures) Recommendations Q.1203 and Q.1213 also refer to these as *Logic Start* and *Logic End*. To keep the terminology consistent, we use only the adjective *Logical* in the book.

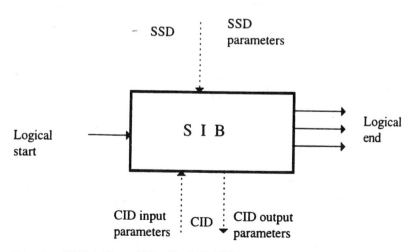

Figure 4.4   SIB interfaces. (*After Fig. 3/Q.1203.*)

instructions, SIBs have parameters; however, drawing further analogies with those would require going into much more detail on computer architecture than the size and purpose of this book warrant.

The dotted lines of Fig. 4.4 demonstrate the data flow in and out of SIBs. These data are categorized into two types: Service Support Data (SSD) and Call Instance Data (CID).

4.2.4.1 Call Instance Data (CID). The CID are the data specific to a particular call instance. Among the examples of such data given in Recommendation Q.1203 are Calling Line Identification (CLI), Personal Identification Number (PIN) entered by a user while making a call, and Called Number.

4.2.4.2 Service Support Data (SSD). The SSD are the data specific to a given service feature and are invariant as far as all call instances are concerned. One specific type of SSD parameter defined in Recommendation Q.1203 is CID Field Pointer (CIDFP), which is a reference to a relevant CID datum. The name and location of a CIDFP-type parameter remains invariant within a given service feature, but its value changes with each call instance.

4.3 Recommendation Q.1213, Intelligent Network—Global Functional Plane for CS-1

4.3.1 Summary

Recommendation Q.1213 was originally published by ITU-T in March 1993, at which time it had 33 pages. The next (CS-1 Refined) version of this document was approved in May 1995. The material of this section is based on the final draft text (Hu, 1995), which has 50 pages.

The main purpose of Recommendation Q.1213 is to define the CS-1 SIBs; in addition, it defines the BCP for CS-1 by specifying its POIs and PORs. To this end, CS-1 specifies 14 SIBs,[51] nine POIs and six PORs. In addition, one SIB, called *Queue,* comes with its Stage 1 SDL description.[52] The rest of this section discusses the BCP and each SIB.

4.3.2 BCP

For a designer of IN-supported services the capabilities to be implemented within the service logic naturally depend on the availability

---

[51]The 1993 version of CS-1 specified 13 SIBs; one more SIB, *Authenticate,* was added as the result of the refinement process.

[52]This SIB implements to the network queuing feature [used, for example, by Freephone (800) service]. This feature is perhaps the most challenging to implementors. For this reason it was paid much attention during standardization. Also for this reason, we follow up on this feature and its implementation through all chapters of this book.

and the specifics of the states of the call processing in which the service logic is invoked. At one extreme, invoking the service logic only when the call has been completed does not leave the service logic much to do. At the other extreme, the ability of the service logic programmer to activate a trigger (or, using software terminology, insert a breakpoint) anywhere in the BCP may easily lead to havoc. What is needed is a sufficient set of well-defined—and safe, as far as the handling of real-time issues and coordination with other activities, such as billing and network management, are concerned—POIs (i.e., the call states where such triggers can be activated), and this is precisely what service designers expect to come out of standardization of the BCP.

Although Recommendation Q.1213 gives several examples of when the same POI can be used for both originating and terminating service features, the specification of POIs does not mention whether a given POI corresponds to the originating or terminating part of the BCP. Of course, there is a good reason for not doing so: the IN Global Functional plane views the BCP as a monolithic process.[53] Nevertheless, it is important to note (and we do that in the discussion of the POIs) that certain POIs may be encountered on behalf of two types of service features (since in most calls there are two parties—the calling party and the called party). But as an effect of the single-endedness principle of CS-1 (described in Chap. 3), one BCP may not have two separate POIs for the originating and terminating service features; instead, two different types of service features can be alternatively assigned to relevant POIs.

To this end, CS-1 has defined the following POIs:

1. *Call Originated.*   This POI corresponds to a state when an end user signals to the network an intention to make a call (e.g., an "off-hook" event before the number is dialed). As far as fixed ("wireline") networks are concerned, this POI may be applicable to, for example, the *Voice-Line*[54] service; depending on what type of signaling is involved in a wireless counterpart of the "off-hook," a wireless application may potentially start just at this point.

2. *Address Collected.*   The local (but not toll) switches in fixed networks collect and process dialed numbers digit by digit. With this procedure, after a certain string of digits has been collected, it is analyzed (at the next POI described in this list) to determine what has to be done next, which may as well be going back to collect more digits. As far as

---

[53]The detailed modeling of the BCP (in which the originating and terminating sides are separated) is considered in the DFP.

[54]A Voice-Line service subscriber speaks (rather than dials) his or her number; the subscriber may also use commands like "Call Mom!" in place of the traditional telephone number.

fixed networks are concerned, there are still debates on whether this POI is useful to IN, but in a wireless network, the whole dialed string of digits corresponding to the address is received and analyzed at the same time, at which point the appropriate service can be invoked.

3. *Address Analyzed.*   Again, in the fixed networks, this is an essential POI for IN. For example, when the prefix 800 is recognized for the Freephone service in the United States, the rest of the number may be used by the switch to determine the appropriate database in the network to send a query to. For the wireless services, this POI coincides with that of Address Collected.

4. *Prepared to Complete Call.*   This POI corresponds to the state of BCP when it is ready to dial the called party. The POI can be used on both the originating and terminating sides, but its most interesting service applications are on the terminating side. There, the POI may be used by CS-1 services like FMD (Follow-Me Diversion) or CF (Call Forwarding). The service subscriber could request that (while his or her own telephone number stays unchanged) the incoming calls be redirected to another number. The terminating exchange (which keeps the line information) could launch a respective query at this point. Another example (mentioned in Recommendation Q.1203) of using this POI is for either terminating or originating call screening.

5. *Busy.*   This POI corresponds to the event of the destination line being busy. Again, it can be encountered on both the originating and terminating side. On the originating side it can be used to allow the calling party to call another number without hanging up. (This capability is important in wireless applications, where both the calling party and the network resources are spared additional authentication processing; similarly, this capability can be used in credit card calling, where the calling party and the network resources are spared additional card validation processing). On the terminating side it can be used in Selective Call Forward on Busy/Don't Answer service.

6. *No Answer.*   This POI is similar to that of Busy because the call to the specified number cannot be completed (at least not by using the line that corresponds to the number). The difference is that the called party's line is free in this case, but no one is answering the call. Everything said about the use of the Busy POI also applies to the No Answer POI.

7. *Call Acceptance.*   This POI corresponds to an unusual—at the first glance—situation when the call is active in that both calling and called parties are on line, but the connection between them has not been established. This POI is most applicable to bridging services; its other application may include services in which the called party may chose to either continue the call or disconnect it (e.g., collect calling).

8. *Active State.*   This POI corresponds to the state in which the calling and called parties are connected. In CS-1 it is applicable to the IN

support of different charging scenarios on the originating or terminating side (or both) and to mid-call processing.

9. *End of Call.*   This POI corresponds to the state when either calling or called party has disconnected. Its main application (on either originating or terminating side) is to invoke the service logic that will free the resources used to support the call, complete charging procedures, etc. In addition, on the originating side it can be used to allow the calling party to call another number without hanging up (cf. the examples for the Busy POI).

This completes the description of the POIs. As far as the PORs are concerned, Recommendation Q.1213 defines the following six:

1. *Continue with Existing Data.*   The BCP resumed at this POR receives no new data from the service logic. The definition in the Recommendation states that the BCP "should continue call processing with no modification," which should be interpreted to mean that the POR is the same as the POI at which the GSL was launched.

2. *Proceed with New Data.*   This POR differs from the previous one only in that the service logic has provided the BCP with the new data.

3. *Handle as Transit.*   Returning at this POR has the effect of restarting the call processing at the initial state of the BCP.

4. *Clear Call.*   Returning at this POR has the effect of terminating the BCP.

5. *Enable Call Party Handling.*   Returning at this POR (which coincides with the POI at which the GSL was launched) enables the BCP to deal with multiple parties. This capability is, of course, out of the scope of CS-1. Recommendation Q.1213 advises that this POR be left for further study.

6. *Initiate Call.*   This POR supports an important (but degenerate) call-processing case when there has been no POI at all: it is the GSL that starts the call. Note that with the model of Fig. 4.2 it is very easy to implement this case (the monitor simply calls the BCP subroutine with the Initiate Call POR the first time).

Overall, At their present state of specification detail the PORs are of much less use than the POIs as far as modeling of the BCP is concerned.

Finally, Recommendation Q.1213 describes the SSD and CID associated with the BCP. The former includes the pointers to the values of the latter, which are as follows: Call Reference (a data structure that uniquely identifies the call in the distributed environment), Calling Line Identity, Calling Line Category (e.g., private line, pay phone, or operator), Dialed Number, Destination Number (initially it is the same as the number dialed, but it may change to, for example, a translated number), and Bearer Capabilities (the ISDN bearer capabilities of the line).

The BCP is described using a SIB template, but, again, there is little—if any—benefit in calling the BCP a SIB. It is much better to keep the concepts clear: SIBs are used for composing service logic.

4.3.3 SIBs

The CS-1 SIBs are described in section 5 of Recommendation Q.1213. In addition, Recommendation Q.1214, which is discussed in the next chapter, provides more information on the use and implementation of all 14 SIBs. As far as Recommendation Q.1213 is concerned, each SIB is described according to the following template:

1. Definition
2. Operation
3. Input (which includes SSD and CID specifications)
4. Output (which includes the description of the branches and CID)
5. Graphical representation (which can actually be derived from the above)

Since the purpose of this book is to help the readers of the IN standards rather than repeat the information contained in them, we will describe SIBs briefly, emphasizing their service applications. For more detail (i.e., the description of all SSD and CID parameters), we refer the reader to the Recommendation itself, whose text on this subject is straightforward.

Finally, we should make one disclaimer. The Recommendation actually specifies that the names of SIBs are to be spelled in capital letters. We find that this requirement is essential in situations that may lead to misunderstandings,[55] for which reason we made an effort to avoid all such situations in the material of this book. For esthetic reasons, we then decided to avoid using all capitals to spell the SIBs' names, although we always capitalize the first letter in each name. We also use the italic font in such cases.

4.3.3.1 Algorithm SIB. Initially (prior to the CS-1 refinement process), this SIB was designed to perform any computation; however, implementation experience dictated that the scope of the computation (and, consequently, the service applications of the SIB) must be narrowed down to be specific. As a result, the "algorithm" performed by this SIB

---

[55]For example, when a reference to a queue, a limit, or an algorithm may be confused with a reference to a name of a SIB (i.e., QUEUE, LIMIT, or ALGORITHM).

is either incrementing or decrementing the value of an integer call-specific variable by a given integer number.

There are two Logical Ends of this SIB: Success and Error. (The latter branch is applicable only to testing the service logic.)

This SIB is applicable to Mass Calling and Televoting services.[56]

4.3.3.2 Authenticate SIB. As its name implies, the *Authenticate* SIB provides the authentication function (i.e., ascertaining that the potential user of a service is a valid one). The SIB permits three options to be implemented: (1) a "simple mechanism," which involves only checking the user's name and password; (2) an "external" mechanism, which involves execution of some externally supplied (presumably via a pointer to a corresponding routine) authentication mechanism; and (3) no authentication at all. (The latter option is not mocking, as it may seem; it helps in reusing service logic. One service provider may wish to use authentication while the other may choose otherwise. In addition, two services may be identical as far their service logic is concerned, but one would be free to a caller and the other would not. Thus, the latter service may require authentication, but the former one would not.) This SIB has two Logical Ends: Error and Success.

Recommendation Q.1213 states that this SIB is applicable to all services. This is true, but it is worthy of notice that it is required by certain services, perhaps the most important example being Personal Communication Services.

4.3.3.3 Charge SIB. The *Charge* SIB is the only one involved with the most serious business issue—collection of the revenue. For this reason alone it has been repeatedly discussed in standards bodies.[57] The charging performed by this SIB is additional to that performed by the BCP.

Recommendation Q.1213 lists the following disclaimers:

- The standard does not define the output format.

- The standard does not necessarily define all information that may be required for charging.

- The standard is not responsible for the subscriber's billing process.

The IN resources to be charged for include circuits, messages (on a packet-by-packet basis), the announcements and storage of voice messages, and the use of the service control computing resources (in time

---

[56]Both services are useful to radio and television stations as they help to evaluate the number of those who follow the broadcast.

[57]In fact, the introduction of parallel execution of SIBs in CS-2 (please see the next section) has been dictated by the observation that charging ought to be performed simultaneously (rather than sequentially) with the action that is to be charged for.

units). The charge itself may be directed to a specific account (including that of the calling party's line or credit card) or a pay phone. The SIB is flexible in that it allows a given account to be charged for only a certain percentage of the overall cost of service provision.

The SIB has two Logical Ends: Success and Error. (The latter branch is applicable only to testing the service logic.)

4.3.3.4   Compare SIB.   This SIB is the service creation counterpart of the assembly-level *COMPARE* instruction reviewed in the beginning of this chapter. The SIB compares a variable of any type with that supplied by the reference value. Depending on the outcome of the comparison, the flow control of the service logic is directed toward one of the following Logical Ends: Greater than, Less than, Equal to, and Error. The latter Logical End is applicable only in testing the service logic.

The *Compare* SIB is polymorphic in that its input may take another form: instead of specifying a variable to be compared with the reference value, a programmer can indicate that the current (to the execution of the SIB) network time be used for the comparison. The network time can be expressed as time of day, day of the week, or day of the year.

Recommendation Q.1213 mentions two service applications: Time-Dependent Routing (a generic feature) and Call Completion to Busy Subscriber. The former allows use of the same number for calling different locations. To appreciate its importance, we provide the classical (and, historically first) example of its use in implementing the AT&T Advanced 800 (Freephone) service feature. Using geographic time differences to its advantage, a service subscriber may choose to keep agents working around the clock in such a way that no agent works very late at night. In addition, in order to ease the network overload, the service subscriber may divert the calls at peak hours to agents located outside of overloaded areas.[58]

4.3.3.5   Distribution SIB.   Similarly to the *Compare* SIB, this SIB can be used to implement the Advanced 800 service features; however, the *Distribution* SIB is a powerful generalization of the Compare SIB. The *Distribution* SIB first executes a specified algorithm (as described below) to determine a Logical End at which the service logic should continue, and then directs the flow of the service logic toward this Logical End. The number of Logical Ends depends on the algorithm type, of which there are four: Time of Day, Day of Week, Percentage, and Sequential. The latter two types require some explanation.

We will again use as an example the AT&T Advanced 800 service features. One such feature provides for incoming calls to be distributed to

---

[58]This application is within the realm of Network Management. In fact, this is an important (and, so far, rare) example of the use of IN in Network Management.

agents on a percentage basis. For instance, out of the three agent locations involved, the first location should get, say, 30 percent of all calls, the second one 27 percent, and, the third one 43 percent. The service logic can implement this feature using the Distribution SIB, which, in this case, is shown in Fig. 4.5.

Note that whenever this SIB is invoked (in the context of a call), it chooses a branch only once. Since the required distribution is maintained on a per-service basis, the SIB must keep certain state information. For example, if a pseudorandom number generator is employed to compute the distribution, the current value of the seed of the distribution has to be kept on behalf of the service. Using the object-oriented terminology, this SIB defines a *class*. When a service is invoked, an *object* of this class is created, and its state is maintained for the duration of the service. In addition, if the SIB with the same algorithm type is to be used in another service feature within the same service, it may require a separate object instantiation depending on the service need.

As far as the Sequential algorithm type is concerned, it simply implies that the SIB selects its Logical Ends in the round-robin fashion. Again, this means that the state of the service execution has to be maintained outside of the call context.

The SIB has an additional Logical End, labeled Error, with the accompanying CID output identifying the error cause. The causes listed in the standard (e.g., invalid parameters' values) suggest that the error handling is useful only for testing the service logic.

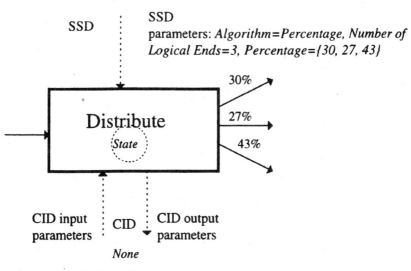

Figure 4.5 Distribution SIB.

4.3.3.6 Limit SIB.   Certain service features[59] require that only a certain number of calls out of all that are dialed get the IN treatment. This is done to *prevent* (rather than respond to) overload of the network resources and is usually carried out in one of the following two ways. If a service[60] is expected to receive P calls every second, but the service processors can handle only $N<P$ calls a second, then only N calls should be presented for processing. Alternatively, the frequency of calls can be expected to be such that the number of calls received within S seconds will fill the service processors for $Q>S$ seconds, which means that for only S seconds out of Q seconds the calls should be given the IN treatment.

The Limit SIB is designed to perform this function. Given the algorithm (via the SSD) to determine the criteria on whether a present call is to be given (or denied) the IN treatment, the SIB executes the algorithm and passes the control to the SIB connected to either of its two Logical Ends, respectively labeled Pass and No Pass. Two algorithm types (described in the previous paragraph) are defined:

1. Pass N calls out of P calls.

2. Pass calls for S seconds out of Q seconds.

Similarly to the *Distribution* SIB, the *Limit* SIB also requires maintenance of the per-service or per-feature (or both) state information. Recommendation Q.1213 specifies that this information be carried in the SSD.

The SIB has an additional Logical End, labeled Error, with the accompanying CID output identifying the error cause. The causes listed in the standard (e.g., invalid parameters' values) suggest that the error handling is useful only for testing the service logic.

Clause 5.8.1 of Recommendation Q.1213 notes that "this SIB is not used for network congestion management function." There are two aspects to the interpretation of this note. First, the SIB deals with the overload of computing resources rather than the trunks or lines of which networks are composed.[61] Second, the SIB is used to prevent—not respond to—the overload.

4.3.3.7 Log Call Information SIB.   This SIB, applicable to all services, logs the call-related information into a file. The SSD include the file

---

[59]Recommendation Q.1213 specifically mentions Time-dependent Call Limiting, although this name is not used in Recommendation Q.1211, which uses the name Call Limiter instead.

[60]Recommendation Q.1213 mentions the following service applications: Mass Calling, Televoting, Freephone.

[61]It is the inability to find a line or trunk to establish a call that manifests network congestion.

name of the file in which the information is to be written, as well as the list of CID pointers to the data to be logged, which may include the call attempt, connect, and disconnect times; calling line identification; dialed number and additional dialed numbers in response to the network prompts (such as credit and calling card numbers and selected options); destination number; bearer capability of the calling line; and, as Recommendation Q.1213 says, any other CID.

The SIB has two Logical Ends, labeled Success and Error. The latter Logical End, as has been the case with all previously described SIBs, is reserved for error handling, which, judging from the CID output error causes, is applicable only to testing. Nevertheless, for this SIB, implementers should foresee additional use of this Logical End. There may be run-time exceptions (e.g., the device on which the file is to be written is out of space) that can be easily remedied (e.g., write the file on another device) by appropriately programming the error path.

4.3.3.8 Queue SIB. The capability to queue incoming calls when all agents are busy has been the major industry requirement that became a primary driver of Advanced 800 service. The requirement was natural. Every time a customer encounters a busy call, the customer is less likely to call back. But if a customer is politely asked to wait for the next service representative, the customer will most likely wait. While waiting, the customer may be entertained by music or given some information about the products and special promotions. Of course, this capability can be (and has been) reused by many other services, which is why it has been standardized as a SIB.

The *Queue* SIB provides the capability to direct the call to a free resource (line or trunk) if it is available or otherwise suspend the service logic processing while waiting for the first resource in a group to become free and then direct the call to this resource. To appreciate the complexity of the requirement, consider the following example.

An 800 Service subscriber has, say, 18 agents to answer calls. Seven of them are located in Chicago, with the remaining 11 in Philadelphia. Each agent has a line identified by the agent's telephone number. It is assumed (and required of the agents) that

1. Only service-directed calls be placed to the agents

2. Agents originate no calls by themselves[62]

Note that, at this point, we intentionally depart—for a moment—from the GFP discipline and consider a simplified version of the Physical Plane. We are not alone in doing so: reflecting on the techni-

---

[62]The technical need for these (otherwise reasonable) requirements will become clear very soon.

cal significance and complexity of the implementation of this SIB, Recommendation Q.1213 provides its Stage 1 SDL description.[63] In order to help the reader understand the description (contained in Fig. 4.6, which is a copy of Fig. 11 of Recommendation Q.1213), we have to descend into the world of messages exchanged between the switches and service control. Of course, we keep the discussion here as schematic as possible;[64] Chaps. 5 and 6 provide a comprehensive treatment of call queuing.

The 800 number in our example is available throughout the country, so it can be dialed from the switching exchanges in Seattle, Los Angeles, and New York, among other places. These switching exchanges have access to the computer that provides service control by executing Global Service Logic. Whenever the 800 call originates at one of the switching exchanges, this exchange places a query to the service control requesting a routing instruction. At this point, the Global Service Logic gets executed by a process created on behalf of the query; this process eventually executes the Queue SIB.

Initially, all of the resources are free, and, so all the resources are marked *available*. While this situation lasts, each service process immediately exits this SIB with the address of the available resource in hand. This address is then passed back to the switch that originated the query. At the same time, the SIB marks the resource *busy*. Assume that the calls arrive at a rate of 20 a minute, while it takes on average three minutes to process a call. That means that in less than a minute after the service becomes operational all agents will be temporarily busy.

This is the situation handled in the SDL diagram of Fig. 4.6. Skipping the parameter checking, etc., the process that executes the SIB takes the following steps:

1. Marks the call as queued.

2. Starts the queuing timer. After all, a call cannot be queued indefinitely (a few of us have encountered an apologetic announcement— followed by disconnecting the call—asking us to call later).

3. Increments the queue count. Again, only so many calls can be queued.

4. Optionally, starts playing announcements to the calling party.

---

[63]No other CS-1 SIB gets this treatment.

[64]To this end, we oversimplify the network issues by considering only the events in the Inter-Exchange Carrier's (IEC's) network. In reality the calls come into the network through the Local Exchange Carriers' (LECs') switches; the agents' terminals may be connected either to the LECs' switches or Public Branch Exchanges (PBXs). The latter may, in turn, be connected via trunks to either LECs' or IEC's switches.

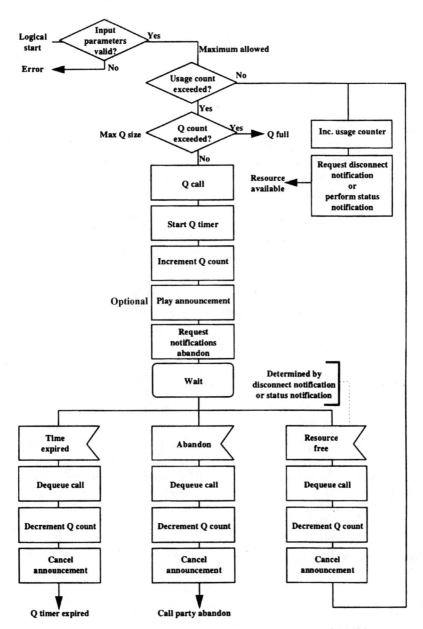

Figure 4.6 Stage 1 SDL diagram of the *Queue* SIB. (*After Fig. 11 Q.1213.*)

5. Requests notification of the call abandonment by the calling party. This message is sent to the originating switch. In this case, the switch will send the message back, if the calling party hangs up. Thus, the network will be able to free the resources.

6. Waits. At this point the process gets suspended until one of the following three events takes place:

   - *The timer is expired.* In this case, everything connected with processing the call (announcements, etc.) is eliminated, and the data structures are reinitialized. The apologetic announcement mentioned in item 2 above may also be played here before the call is disconnected.

   - *The call is abandoned.* In this case, again, all resources that were held on behalf of the call are released.

   - *A resource is free.* This is the most interesting case. First of all, how did the service logic learn that it became free? Dramatically, the answer to this question lies in the way the event was processed. After dequeuing the call and cleaning up the data structures, the originating switch is handed both the address of the freed line and the request for the disconnect notification. When the calling party is through with the call, the originating switch will send this notification to service control, which will manifest the event that this particular resource is free. In other words, the call that is handed a resource makes sure (by requesting the disconnect notification from the switch) that when it terminates (thus releasing the resource), the queue will advance.

Note again that all the messages that we considered for the purpose of our discussion are exchanged between the service control and the originating switch. None of the terminating ones (to which the lines are actually attached) have been involved in the IN processing! If that had been the case, many more messages would have been needed, and, what is worse, these messages would spread through the network, tremendously increasing its load and the cost of operating it. The only price of keeping the communications to the originating switch was the requirement that only service-related calls be directed to the agents and that agents may not originate calls (otherwise, service control, which marks resources busy only when it hands them to a call, could be mistaken about the their states).[65]

The SSD of this SIB specifies the limits in the queue length, number of active calls, and the maximum time in queue, as well as the set of

---

[65]This could, of course be remedied by using *continuous monitoring* (please see the description of the Status Notification SIB in Sec. 4.3.3.11), but this would be an expensive solution, which could also result in SS No. 7 network overload.

resources for which the calls are to be queued.[66] In addition, it contains the same announcement information as the User Interaction SIB is supplied with, for which we refer the reader to Sec. 4.3.3.13. The CID input for this SIB contains the call reference.

The SIB has five Logical Ends, respectively labeled Resource Available, Call Party Abandon, Queue Timer Expir[ation], Queue Full, and Error. The latter Logical End corresponds to programming errors and is useful only for testing purposes.

4.3.3.9 Screen SIB. This SIB determines whether a given string [e.g., Personal Identification Number (PIN), originating or terminating number] is contained in a stored list of strings, called *screen list*. The SIB is applicable to all services that require terminating and originating call screening as well as those that require verification of user-entered data (e.g., Credit Card Calling and Account Card Calling).

The SSD of *Screen* SIB contain the screen list and the screen list filter, which identifies additional procedures to be used during screening. The CID of this SIB contain the filtering parameters and an attribute called Authorized Relationship Id. The latter is necessary for the security reasons: the computer on which the screening is performed is part of a distributed system, and the screening transaction requires communications with other computers. To make sure that these communications take place between authorized parties, the transactions (or *relationships*) among the parties are strictly monitored.

There are two Logical Ends, Match and No Match, which, respectively, correspond to the successful and unsuccessful outcomes of screening and an additional Logical End, Error, with the accompanying CID output identifying the error cause (invalid parameters' values).

4.3.3.10 Service Data Management SIB. This SIB provides the capability to manipulate the user-specific data in the network database. Such manipulations can be performed by the service management (rather than service control) software,[67] but there are also services (e.g., Call Forwarding, Universal Personal Telecommunications, and Wireless Personal Communications Services) where the service logic has to update users' profiles.

The actions on the users' data performed by the Service Data Management SIB support

1. Adding objects to and removing them from the database

---

[66]The latter item was omitted from the description of the SSD in Recommendation Q.1213 and, mistakenly, placed in the CID. This mistake will be fixed in CS-2.

[67]In fact, this SIB is the only one that explicitly performs service management functions.

2. Retrieving and changing the values of the attributes of the objects in the database

3. Resetting the values of the attributes to their defaults

4. Incrementing and decrementing the values of the attributes by specified amounts

The SSD data contain the pointers to the objects in the database. The CID data specify the action to be performed, the object name, the attribute name (for all but the addObject and removeObject actions), and the amount used for incrementing or decrementing attribute values (for Increment and Decrement) actions. The CID output data contain the retrieved values (in case of the Retrieve action) and—in case of a failure to perform a specified action—the error cause (e.g., invalid action or value, etc.).

The Service Data Management SIB has two Logical Ends, labeled Success and Error. Note that in the case of this SIB, the Error Logical End is highly relevant to the run-time recovery of service logic: if the input received from a user (via the User Interaction SIB) is invalid, the Error branch may, for example, lead to another user interaction session, in which the error can be explained to the user. The user could then be given another opportunity to enter the data.

4.3.3.11 Status Notification SIB. This SIB provides the capability to monitor the status of particular network resources (here meaning either lines or trunks whose status may be either *idle* or *busy*). The Status Notification SIB is applicable to the following services: Completion of Call to Busy Subscriber, Freephone, and Call Transfer. The four status notification capabilities defined in Recommendation Q.1213 are as follows:

1. *Polling* (i.e., checking the current status of the specified resource).

2. *Waiting* (but for no longer than a specified time limit) until a specified resource changes its current status (either from *busy* to *idle* or vice versa, although the latter case does not seem to have an obvious service application). The invocation of the SIB results in suspending a process in whose context this SIB was executed.

3. *Initiating continuous monitoring.* This demands significant distributed activity, for it requires, unlike any of the other two capabilities, that an additional process be started by the network. This process, which is to run for a specified time period, keeps track of every status change of a specified resource and possibly records it in the resource history file. The process in whose context the SIB was executed continues concurrently with the latter process.

4. *Canceling continuous monitoring.*

The SSD include the specification of one of the above status notification types, the resource, and other parameters required by the status notification type selected (e.g., the name of the resource history file for the continuous monitoring). Note that all but the Wait for Status type are nonwaiting instructions: the service logic process continues regardless of any external event having taken (or not taken) place. With Wait for Status, however, the service logic process is suspended until a certain external event (change of status of the resource monitored) occurs. If the event never occurs (or if its occurrence is not properly reported to the service logic process), the service logic process will hang forever. To prevent this from happening, a timer is started as waiting begins. The value of the timer is specified in the SSD, and, if the waiting time exceeds this value, an *internal* event generated by the timer will get the service logic out of the stupor.

The three Logical Ends of this SIB are labeled Success, Timer Expir[ation], and Error. Error is useful for testing (the error causes are limited to identification of wrong parameters). Timer Expir signals an important exception as explained in the previous paragraph. Finally, Success leads to the branch at which the service logic should continue normal processing.

4.3.3.12 Translate SIB. This SIB provides the core IN capability, as required by the key benchmark services (Freephone, Virtual Private Networks, Universal Personal Telecommunications, etc.). The combination of the input information and CID (e.g., calling line identification) is translated into a network address. Clause 5.14.2 of Recommendation Q.1213 gives an example of using this SIB for modifying the input information into "a standard numbering plan upon which the network routing is based."

The SSD of the *Translate* SIB include the pointer to the input data object, the filter (whose role is analogous to filtering performed by the Screen SIB), and the pointer to the translated data. The CID contain the filter values and the Authorized Relationship Id value.

The SIB has two Logical Ends, Success and Error. The latter, in addition to signaling parameter value problems (specified in the CID output), may designate the condition where translation is not available (due to problems with external databases, for example). This latter case allows for run-time exception handling.

4.3.3.13 User Interaction SIB. An essential part of IN is the ability to play announcements to end users and solicit their input[68] in order to determine where to direct the call. In fact, in many services, multiple

---

[68]Until recently, this input took the form of dialed multifrequency tones, but, with the integration of voice recognition into the network equipment, users are more and more often prompted to speak rather than dial.

interactions take place before the call is finally connected to the called party; in some cases,[69] there may be no other activity in the call.

The User Interaction SIB provides the capability to exchange information between a network and an end user.[70] Recommendation Q.1213 declares that "most IN CS-1 services will use this SIB."

The SSD hold the whole range of the announcement parameters, which specify

1. The announcement to be played.

2. Whether this announcement is to be repeated (and, if so, the interannouncement interval in seconds, the maximum number of repetitions, and the maximum time to be spent playing the announcement.[71]

3. Whether any data are to be collected from the user, and, if so, the format of the data (e.g., audio or a string of characters) and whether the announcement is interruptible (i.e., the user may speak or enter digits before the announcement is completed). In addition, if the format of the data to be collected is a character string, the parameters would include the minimum and maximum sizes of the string, intercharacter waiting time,[72] and, if needed, the input delimiter character.

The CID input identifies the call party; the CID output holds the collected data or an error cause. The SIB has two Logical Ends, Success (associated with the collected data) and Error (associated with the

---

[69]Consider a service subscribed to by a company that sells items advertised in a catalogue. When a customer calls the service, he or she is prompted by the network to enter the item number and the credit card number. Those are recorded (together with the customer's address, which may be determined from the originating telephone number or obtained from the customer by additional prompting). The company then ships the purchased item to the customer.

[70]The same capability is partly supported by the Queue SIB. Even though it would be better to keep user interaction capabilities in one place, the problem is that in CS-1 it is impossible to invoke a SIB inside of another SIB; nor is it possible to allow two SIBs to execute concurrently. This problem will be remedied by CS-2, which will support both constructs.

[71]This parameter may appear redundant, since, given the announcement time t, the maximum number of repetitions R, and the interannouncement delay d, the maximum time spent playing announcements T could be calculated as follows: $T = R(t + d) - d$. If set consistently, the maximum time parameter may control only the duration of the *last* announcement. It should be noted, however, that announcements may be charged for on a second-by-second basis, and a single announcement can be long enough to be expensive. It is for this reason that the maximum time parameter has been introduced.

[72]This not only eliminates endless waiting, but also provides a natural end-of-input delimiter: If nothing is entered after a specified period of time, the input is considered complete.

error cause). The error conditions for this SIB include wrong parameters, wrong call state, and user-dependent run-time conditions (call abandonment, collection time-out, incorrect number of characters, and unavailability of announcement resource). The first two conditions correspond to system errors, which presumably ought to be fixed before the service has been deployed, but the third condition—or rather set of conditions—is essential for the proper function of the service.

The support of this SIB by the network is quite elaborate. Both the Distributed Functional Plane and Physical Plane contain various procedural scenarios and several options as far as the related distributed processing is concerned.

4.3.3.14   Verify SIB.   Clause 5.16.1 of Recommendation Q.1213 suggests that "the *Verify* SIB normally follows the *User Interaction* SIB when information has been collected from a call party." Therefore this SIB is applicable to the same services that employ the *User Interaction* SIB. The role of the SIB is to parse the input data to determine whether they are syntactically correct.

To this end, any grammar can be ultimately specified by the SSD, but, for the time being, it is limited to describing the alphanumeric string expressions (delimited by characters "#" or "*") of limited length and rather simple format.

The Verify SIB has three Logical Ends, Pass (and indicating the positive outcome of parsing), Fail, and Error. In case of the latter, the accompanying CID output identifies the error cause (invalid parameters' values), which makes this Logical End specifically applicable to testing service logic.

4.4   Draft Recommendation Q.1223,
Intelligent Network—Global Functional Plane
for CS-2 (Highlights)

4.4.1   Overview

The version of the Draft Recommendation[73] available at the time of writing this book is 90 pages long. The Recommendation has introduced the following new modeling constructs and capabilities (which are described under their respective headings in the rest of this section):

- Basic-Call-Unrelated Process
- SIB operations

---

[73]See Hu and Herian (1996).

- High-Level SIBs (HLSIBs)
- Parallel Processing
- Call Party Handling

As far as SIBs are concerned, the Recommendation now defines 21 of them. The CS-1 Limit SIB has been taken out, and eight new SIBs[74] have been introduced. Note, however, that the definition of SIB has been changed (see Sec. 4.4.3); if the old terminology had been retained, there would have been even more SIBs—as many as there are SIB *operations* now. The new SIBs are described in the context of relevant topics except for two SIBs, Service Filter and User Interaction Script, which do not seem to have stable definitions.

The present description of the BCP is also incomplete, awaiting the final decisions concerning the CS-2 Call Model. Overall, the reader should keep in mind that the present material may differ from what is published in the final version of the Recommendation.

4.4.2   Basic-Call-Unrelated Process

The Basic-Call-Unrelated Process (BCUP) is the counterpart of the BCP for modeling the capabilities implemented through actions that are *not* performed on behalf of a particular call (even though they are necessary for supporting calls). A few examples of such capabilities[75] are user authentication, user registration, user screening, user activation, and user deactivation. (These capabilities are required by Personal Communications Services.) As in the case of the BCP, the processing of the BCUP may be suspended and the external service logic may be invoked, upon the execution of which the BCUP will continue.

As far as the interfaces with the GSL are concerned, two POIs and three PORs have been presently defined for the BCUP. The POIs are Analyzed Message (to support the Location Updating feature of wireless PCS) and Answer (to signal that the connection between the network and the user has been established). The PORs are Continue with Existing Data, Proceed with New Data, and Release Association. The first two PORs have the same meaning as in the case of the BCP; the last one instructs the BCUP to release the association between the network and the user.

---

[74]These SIBs are Join, Service Filter, Split, Initiate Service Process, Send, Wait, End, and User Interaction Script.

[75]Given in clause 6.2.3.3 of the present text.

4.4.3 SIB operations

The Recommendation introduces a new term, *SIB operation,* described (in the present clause 4) as the "non-interruptible and atomic function performed within a SIB relating to the SIB's capability." With this definition, SIB operations (rather than SIBs, which are no longer atomic) correspond to assembly-level instruction, as they are observed by a programmer. The old term (i.e., *SIB*) is reserved for designating the "network capability consisting of [executing] the SIB operations." Compared to the CS-1 SIBs, CS-2 SIBs improve the granularity of the programming capabilities. The naming change is quite radical, however, since the word "SIB" now denotes a library of instructions rather than a single instruction.[76]

There may be several SIB operations related to the same SIB, and, in fact, several CS-1 SIBs have been expanded to contain more than one operation. The following example of the new Queue SIB should well illustrate the new concepts.

This Queue SIB now contains two SIB operations: Queue Call and Queue Monitor.

The Queue Call SIB operation has five *outlets*[77] in its Logical End, which are respectively labeled Resource Available, Queue Full, Call Party Abandon, Call Queued, and Error. If the call was queued, the SIB would *immediately* return control at its Call Queued outlet.[78] At this point, service programmers are given multiple possibilities.

One of them is to attach to the second SIB operation, Queue Monitor, to this outlet. The *Queue Monitor* SIB operation also has five outlets in its Logical End, which are respectively labeled Resource Available, Queue Timer Expiry, Call Party Abandon, Message, and Error. Of these, the penultimate one is new (compared to CS-1). When connected to the User Interaction SIB operation, this outlet should result in the execution of the User Interaction SIB for as long as the call is kept in the queue.

---

[76]Perhaps it would have been more consistent with the CS-1 terminology, if what is now called *SIBs* had been called *capabilities,* and the *SIB operations* would still have been called *SIBs.* Nevertheless, the present definition may help in the future work of aligning SIBs with the Application Service Elements (ASEs) of the Physical Plane.

[77]This is another change in the CS-2 definitions. The term *Logical End* is now reserved to denote a combination of *all outlets* (formerly, Logical Ends).

[78]In other words, the Queue Call SIB operation is what is called a *nonwaited instruction,* unlike the CS-1 Queue SIB, which in a similar case would have resulted in a suspension of the service logic until the call was dequeued.

4.4.4 High-Level SIBs (HLSIBs)

The Recommendation makes the next necessary step in extending service programmability by introducing the constructs that allow definition of new SIB operations by using the existing ones. Thus, High-Level SIBs (HLSIBs) are introduced as the entities "composed of SIB operations and/or other SIBs." The formal composition process is not specified in the Recommendation, but an example given there alludes to the macro mechanism that has been in use in Assembly languages for many years. Namely, an instruction to be introduced is first described in a macro as a sequence of existing instructions. Whenever such an instruction is encountered in the text of the program, it is substituted for by the sequence in such a way that the names of the operands (i.e., the SSD, in the case of SIB operations) supplied with the instruction are consistently substituted throughout all references to them by the existing instructions. In terms of the graphical SIB representations, that simply means that the SSD of an HLSIB are mapped into those in its defining chain. Thus, if the SSD parameter name of a SIB in the defining chain does not match any parameter name of the HLSIB, this parameter is considered "local." The interpretation of this is left open for now; most likely, it means that the value should be explicitly assigned to it within the service logic program.

4.4.5 Parallel processing

An essential new feature of the model is its support of parallel execution of GSL; that is, one service logic process may spawn another one to run concurrently with its own execution. The point in the execution of the former process at which the new process is created retains the name of Point of Initiation (POI).[79]

Whenever parallel processing is introduced, it is essential to define the capabilities of the processes to exchange messages as well as the *discipline* guarding the exchange. Recommendation Q.1223 states that service logic processes may communicate via CID;[80] in addition, the Recommendation describes the synchronization mechanism by which each process can have a set of preprogrammed *Points of Synchronization* (POSs) at which it has to stop and wait for a message from another process. If the sending process were also required to wait until the receiving process actually receives the message it was sent, the

---

[79]Hence, this term now applies not only to the BCP but to service logic processes as well.

[80]Note, however, that no data access mechanisms have been so far specified for accessing the (shared) CID.

communications discipline proposed by the Recommendation would be what is known as *rendezvous,*[81] which is the simplest mechanism for interprocess communications. But the Recommendation explicitly states that the sending process is allowed to continue after it has issued the message and that the message be queued at the POS of the receiving process.

Four new SIBs, Initiate Service Process, End, Send, and Wait are introduced[82] in support of parallel processing. The first two SIBs are used for respectively starting and terminating a service process; the Send SIB is used for sending a message to another service process; and the Wait SIB is used for establishing a POS.

4.4.6 Call Party Handling[83]

Two new SIBs, Join and Split,[84] have been introduced in the present version of the Recommendation. The invocation of the Join SIB operations results in connecting "two call parties together to establish [a] speech path." The effect of the Split SIB operation is inverse: the speech path to a specified call party is eliminated.

---

[81]See, for example, Tanenbaum (1992) or Peterson and Silberschatz (1985).

[82]Each of these four SIBs contains only one SIB operation of the same name.

[83]These SIB operations have no detailed description in the present version of the recommendation.

IN Distributed Functional Plane

5.1 Overview

While both the Service and Global Functional Planes deal with the *what* of service support, the Distributed Functional Plane is the first place where the *how* issues surface. The word *function* is used here in the sense of a (more modern) term *class* (of *objects*), and the network is viewed as a set of objects, called functional entities (FEs), which exchange messages called *Information Flows* (IFs) over the abstract media called *relationships*.

At the Physical Plane, these objects or their combinations are mapped into (and consequently implemented within) the physical entities (PEs); the relationships are mapped into physical media (via protocol stacks); and the IFs are mapped into appropriate protocol messages. Often the question is asked, why does one need this extra layer of description? In other words, why not deal with the Physical Plane right away, bypassing the present one altogether?

Here is one answer to this question. If there were only one way to implement the FEs within the PEs, the view of the IN Distributed Functional Plane would indeed be superfluous; however, as the next chapter demonstrates, there are several possible mappings. Consider three mappings, which respectively place two FEs into

1. The same PE
2. Two different PEs connected via SS No. 7 Transaction Capabilities (TCAP) protocol

3. Two different PEs connected via SS No. 7 ISDN User Part (ISUP)
   protocol[85]

In the first case, the IFs can be implemented as procedure calls or
interprocess messages (if the FEs are implemented as software modules) or as electronic signals (if the FEs are implemented as hardware
modules). In the second case, the IFs have to be implemented using the
TCAP platform (which, for example, may mean they have to be broken
into smaller components). In the third case, the ISUP protocol itself
may have to be updated. Thus, three different implementations definitely require three different specifications, each of which is to be
derived from scratch. Yet, in all three cases, exactly same information
is to be exchanged in support of the same set of capabilities! By specifying the information exchange on the same level of abstraction as the
FEs are specified, one gets the universal specification, which can later
be implemented on a specific platform.

The exchange of IFs among the FEs is governed by certain rules,
which constitute the abstract IN protocol. More precisely, following the
taxonomy of Holzmann (1991), a protocol specification should explicitly include the following five elements:

1. *Vocabulary* (i.e., the set of protocol messages)

2. *Assumptions* about the environment where the protocol is executed

3. *Service*[86] to be provided by the protocol

4. *Procedures* (i.e., the rules that sequence the messages)

5. *Encoding* of each message

This taxonomy applies to the Distributed Functional Plane,[87] as
follows:

1. The vocabulary of the protocol, as described at the level of the
   Distributed Functional Plane, consists of IFs, each of which carries
   Information Elements (IEs).

---

[85]This switch-to-switch protocol is specified in ITU-T Recommendation Q.762 (1993),
"Signalling System No. 7—General Function of Messages and Signals of the ISDN User
Part of Signalling System No. 7," and ITU-T Recommendation Q.763 (1993), "Signalling
System No. 7—Formats and Codes of the ISDN User Part of Signalling System No. 7." A
good relevant monograph is Stallings (1992). The choice of protocols in this example
(although very real) is not essential for making the point that one abstract specification
can be used by several implementation platforms.

[86]Naturally, the word *service* here has a somewhat different meaning than that used
in previous chapters. In relation to protocol, the service description demonstrates its
purpose and the manner in which it is used.

[87]In the next chapter, we also consider the Physical Plane in view of this taxonomy.

2. The assumptions about the environment are stated in the description of the FEs and the relationships among them.

3. The description of protocol services (i.e., the *semantics* of the protocol) indicates on whose behalf the protocol is executed and what has been achieved as the result of the execution. The Distributed Functional Plane Recommendation addresses this item in several places. Recommendation Q.1214, for example, describes the IFs in the context of each SIB. In addition, for each IF, it explains its role, pre- and postconditions, as well as each of the IEs that this IF carries.

4. The procedures are specified, first of all, through the introduction of the *Basic Call State Model* (BCSM), whose two objects, *originating* and *terminating* calls, spur most of the IN traffic. In addition, both call flows and SDL[88] descriptions are used to describe the sequencing of the messages.

5. The encoding (i.e., representation of the messages as strings of digits)[89] need not be specified because the messages at this level of abstraction are not actually destined to go over physical pipes.[90]

We will further discuss this taxonomy as we proceed in reviewing specific Recommendations. The rest of this chapter reviews Recommendation Q.1204, relevant aspects of Recommendation Q.1211, Recommendation Q.1214, and the highlights of the present status of Draft Recommendation Q.1224, in that order.[91]

5.2   Recommendation Q.1204, Intelligent Network—Distributed Functional Plane

5.2.1   Summary

This Recommendation is 25 pages long. Its initial (and still current) publication is dated March 1993.

The main text of Recommendation Q.1204 first describes the Distributed Functional Plane (DFP) architecture and defines the IN

---

[88]Specification and Description Language (SDL) has been traditionally the language of choice in telecommunications. We have already used an illustrative SDL program in the previous chapter. To fully understand every detail of the IN standard (and most other ITU-T Recommendations), the reader would have to learn SDL. It is not that difficult; we recommend ITU-T Z.100 (1994) as a reference and Olsen et al. (1994) as a comprehensive monograph.

[89]Most often, binary digits (0 and 1) are used for this purpose.

[90]Cf. the Physical Plane where the Abstract Syntax Notation (ASN.1) language is used in lieu of direct encoding.

[91]The relevant text of Draft Recommendation 1221 has not been developed yet.

FEs and relationships among them. It proceeds by defining the modeling concepts and their use. After that the relationship to the INCM is discussed, especially, the mapping of the SIBs into the Distributed Functional Plane [the SIBs are executed by the FEs as they perform Functional Entity Actions (FEAs) and exchange IFs across their relationships].

The Recommendation has three annexes:[92]

1. Annex A introduces the *Basic Call State Model* (BCSM).

2. Annex B discusses the benefits of object-oriented modeling.

3. Annex C introduces the concepts of the *Call Segment Model* and *Call Views.*

In the rest of this section, we review the three key components of the IN protocol environment: the FEs, their relationships, and the IN call model.

5.2.2 Functional Entities (FEs)

The discussion of FEs is accompanied by Fig. 5.1 (which is a copy of the Fig. 2.1 of Recommendation Q.1204). It should be repeated that the FEs are abstract models of the hardware and software modules used in the network equipment. If the reader feels at any time lost about the relation of the model to the physical world, he or she may consult the part of the next chapter that deals with Recommendation Q.1205, where the relation of FEs to PEs is described.

The IN FEs are grouped according to their role in supporting IN: (1) those that are involved in service execution and (2) those that are involved in service creation and management.

5.2.2.1 Service execution FEs.   The six service execution FEs are as follows:

1. The *Call Control Agent Function* (CCAF) provides user access capabilities. It may be viewed as a terminal through which a user interacts with the network. With this definition, the role of CCAF in IN may appear questionable at first glance. After all, according to IN principles, at the user level, the IN support (or lack of it) of any given service is unrecognizable. Nevertheless, on the network side, it is essential to know the type of the user terminal to provide a proper service. For example, if the CCAF does not have Dual Tone Multi-

---

[92]An annex within an ITU Recommendation forms an integral part of this Recommendation as far as the standards go. Thus, the material of any such annex is *normative.* (On the contrary, the material of an appendix of an ITU Recommendation is *informative.*)

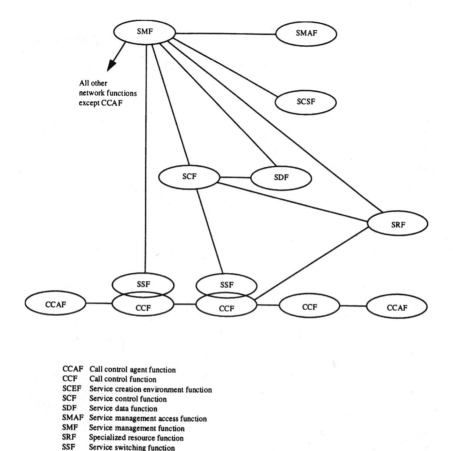

All other
network functions
except CCAF

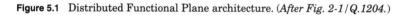

CCAF Call control agent function
CCF   Call control function
SCEF Service creation environment function
SCF   Service control function
SDF   Service data function
SMAF Service management access function
SMF   Service management function
SRF   Specialized resource function
SSF   Service switching function

Figure 5.1 Distributed Functional Plane architecture. (*After Fig. 2-1 / Q.1204.*)

Frequency (DTMF) capability, and the user interaction capability is to be invoked, the service control should be able to determine that either voice-recognition facilities or a human operator must be connected to the user. And, of course, the support of some aspects of mobile telephony requires altogether much different technical effort than that of fixed terminals.

2. The *Call Control Function* (CCF) provides the basic switching capabilities available in any (IN or non-IN) switching system. These include the capabilities to establish, manipulate, and release calls and connections.[93] It is the CCF that provides the trigger capabilities discussed in the previous chapter; however, another object called *Service*

---

[93]It is a modern axiom of telecommunications that services, calls, and connections (that comprise a call) should be separated. IN is making the first step by separating services from calls and connections, but the latter two are still combined within one entity.

Switching Function (SSF) is needed to support the recognition of triggers as well as interactions with the service control. Finally, at the Distributed Functional Plane level, the POI/POR modeling of the call is replaced by a precise Basic Call State Model (BCSM) of CCF, which is reviewed in the next section.

3. The *Service Switching Function* (SSF), as mentioned in the previous paragraph, cooperates with the CCF in recognizing the triggers and interacting with the service control. (An example of the SSF role is suspension of call processing so that service control can convert an 800/Freephone number into an appropriate network address.) Note that Fig. 5.1 depicts the CCF and SSF as overlapping ovals. This is to signal that these objects are inseparable: for now, a network element containing the SSF must, by definition, also contain the CCF. For that matter, the notation SSF/CCF is used throughout all IN Recommendations to refer to a class of objects with switching capabilities.

4. The *Service Control Function* (SCF) executes service logic. It provides capabilities to influence call processing by requesting the SSF/CCF and other service execution FEs to perform specified actions. Implicitly, the SCF provides mechanisms for introducing new services and service features *independent* of switching systems. Historically (see Chap. 1), SCFs started as network databases.

5. The *Specialized Resource Function* (SRF) provides a set of real-time capabilities, which Recommendation Q.1204 calls specialized. These capabilities include playing announcements and collecting user input[94] (either DTMF or voice, depending on the facilities). The SRF is also responsible for conference bridging and certain types of protocol conversion as well as text-to-voice conversion. Recommendation Q.1204 states[95] that the SRF "may contain functionality similar to the CCF to manage bearer connections to the specialized resources."

6. The *Service Data Function* (SDF) provides generic database capabilities to either the SCF or another SDF. It is important to note—and we will soon return to this point—that this capability is completely separated from that of service control. As far as service examples go, the SDF is used to manage the data for

- Cross-network cooperative number translation performed for VPN and PCS

- Screening and authorization, performed on behalf of Account Card Calling service

- Customer record maintenance (relevant to all services)

---

[94]That is, performing the function described in the User Interaction SIB section of the previous chapter.

[95]In Clause 2.4.

5.2.2.2 Service creation and management FEs.   The following three FEs are defined in Recommendation Q.1204:

1. *Service Creation Environment Function* (SCEF), which is responsible for developing (programming) and testing service logic, which is then sent to the service management function (SMF)

2. *Service Management Function* (SMF), which deploys the service logic (originally developed within the SCEF) to the service execution FEs, and otherwise administers these FEs by supplying user-defined parameters for customization of the service and collecting from them the billing information and service execution statistics

3. *Service Management Agent Function* (SMAF), which acts as a terminal that provides the user interface to the SMF[96]

5.2.3 Relationships among FEs

Clause 4.1 of Recommendation Q.1204 defines these relationships based on the client-server model: "In order for one Functional Entity (termed the client Functional Entity) to invoke the capabilities provided by another Functional Entity (termed the server Functional Entity), a relationship must be established between the two Functional Entities concerned.... A relationship between two Functional Entities can only be established by the client Functional Entity.... A relationship between two Functional Entities can be terminated by either Functional Entity." It is also true, and Recommendation Q.1204 makes this point, that under different circumstances *either* of the two FEs involved may play the role of a client (for whom the other one becomes a server). Complex scenarios (such as Service Assist and Hand-off considered later in this book) demonstrate the latter point, but even simple service circumstances can illustrate it.

---

[96]Because of the highly competitive nature of service creation, very little has been said about it in the standards documents. One natural question that arises is why there is a user-interface agent (terminal) for service management (i.e., SMAF) but none for service creation. Although a review of available IN-related products is outside of the scope of this book (for which reason the book intentionally avoids mentioning such products), at this point we observe that some vendors have separate systems for service creation and service management so that each system has a distinct user interface. Other vendors implement a service creation environment *and* a service management environment within the same system, which requires only one user agent function. Although an architecture that supports the *Service Creation Environment Agent Function* (SCEAF) had been accepted by ITU and the augmented architecture had been considered by the industry (Omiya and Suzuki, 1994), this architecture has not yet been reflected in the text of any Recommendation. On the other hand, given that the user-interface-related issues of service creation and service management are out of the scope of standardization, the issue is moot, and neither absence nor presence of a function relevant to the user interface is significant.

Consider a Freephone service. After CCAF passes dialed digits to the SSF/CCF, the latter launches a query to the SCF. In this case, the SSF/CCF is a client and the SCF is a server. On the other hand, the IN implementation of a *Wake Up* service may result in the SCF requesting that the SSF/CCF dial the party that had ordered the wake-up call. In this case, the SCF is a client and the SSF/CCF is a server.

The following relationships are valid:[97]

1. SCF with SSF/CCF, SRF, SDF, and SCF

2. SDF with SDF

3. SMF with SSF/CCF, SRF, SDF, SCF, SCEF, and SMAF

Straight line segments connecting the FEs in Fig. 5.1 represent the valid relationships on the Distributed Functional Plane. Note that one FE may have several relationships with other FEs.

IFs are exchanged across relationships. Each IF is either a client's request or a server's response.

5.2.4   Modeling with finite state machines (FSMs)

Recommendation Q.1204 mandates the use of finite state machines (FSMs) for the description of the protocol-related behavior of the FEs. To this end, FSMs are used in Recommendations Q.1204 and Q.1214 for call modeling. In addition, FSMs are also used in the IN interface Recommendations Q.1218 and Q.1228. The Recommendations assume that the reader of standards is familiar with FSMs.

For this reason, we introduce the FSM technique in the present chapter and urge the reader who is unfamiliar with the technique to become acquainted with it—it is intuitive and reasonably straightforward, and it is central to understanding the IN protocol. Of course, the reader who is familiar with FSMs may skip the rest of this section.

Before giving the formal definition of an FSM, we consider an illustrative example. One interacts with an object by sending messages to it. The object responds to these messages with certain actions. Sometimes, the same message may require a different action depending on what happens with the object at the time it is receiving a message. Consider a telephone terminal, which may receive the following "messages" from either the human user or the network:

- The user is lifting the receiver.

- The user is putting the receiver down.

---

[97]That is, valid *in principle*. Each Capability Set defines its own subset of valid relationships.

- The user is dialing a number by pushing the buttons.

- The network is alerting the terminal that a call is received.

Naturally, if the user tries to dial *before* lifting a receiver, there will be no observable action. After the user has lifted the receiver, dialing will put the user in touch with the network. Thus, we can define two states of a telephone terminal regarding the position of the receiver: Receiver_Down and Receiver_Up. In each state, we can consider the following *events,* which are caused by the messages sent (by either a person or the network) to the terminal: Lifting_Receiver, Putting_Receiver_Down, Dialing, and Alerting. Depending on the state in which these events take place the terminal may act as follows:

- Do nothing (i.e., exhibit no observable, from the modeling point of view, action).

- Send the *Connect*[98] message to the network indicating that the receiver has been lifted.

- Send the *Disconnect* message to the network indicating that the receiver has been put down.

- Send the dialed digits to the network.

- Provide the user with the dial tone.

- Ring the user.

We say that the terminal may produce the following *actions,* which are respectively denoted by the following symbols: *Do_Nothing, Connect_to_Network, Disconnect_to_Network, Send_Digits_to_Network, Provide_Dial_Tone,* and *Ring_User.*

For each state, the model considers the *events* that may take place in it.[99] For example, in our model, when the receiver is up, the network may not alert the terminal (we assume that the network would know that it is busy). If the event may take a place in a state, then a *transition* is specified, which indicates the following two things:

1. The action[100] (or set of actions) to be performed when the event takes place

---

[98]The names of this message and the next one are invented for illustrative purposes. In the case of POTS, there is no "message" other than the appearance of an electronic signal; the situation is, of course, different in the case of ISDN.

[99]In distributed systems, such events are often caused by the reception of *input* messages, for which reason they are sometimes called *inputs.*

[100]In the description of distributed systems, such an action may take form of *internal* processing within a single entity or an *output* message emitted by an entity.

2. The next state the object is to move to[101]

(It is assumed that all actions and the transition to the next state are performed at the same time the event takes place.)

For example, if the system is in the *Receiver_Down* state, and the event is *Dialing,* the action of the terminal is *Do_Nothing,* and the transition leaves the telephone terminal object in the same state. The *Lifting_Receiver* event, however, will cause two actions, *Receiver_ Up_to_Network* and *Provide_Dial_Tone,* [102] and it will also cause a transition to the *Receiver_Up* state.

The set of all states, events, transitions, and actions specifies a given FSM.

We graphically depict each state by a rectangle[103] containing the name of the state. Each transition is depicted by an arrow that connects the two states involved. Each such arrow is marked with the symbol of the event that causes it, followed by the "/" character, followed by the list of symbols that correspond to the actions to be performed. The fully specified FSM for our model of the telephone terminal is depicted in Fig. 5.2. The reader should familiarize himself or herself with this FSM to the point of satisfaction. We also recommend that the reader go through a few exercises by changing the model to make it more realis-

---

[101]This state can be the same the object is presently in.

[102]Of course, in reality, it is the network that provides the dial tone to the terminal after the terminal signals the network that the receiver is up. Thus in a more realistic model, we would add another state, say, Waiting_for_Dial_Tone, and another event.

[103]More often, circles are used for this purpose. The IN Recommendations, however, have been using rectangles.

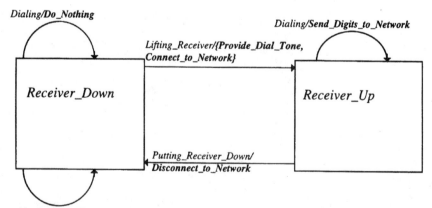

Figure 5.2 Example FSM.

tic (i.e., adding states that correspond to waiting for the dial tone, conversation, etc.,[104] as well as appropriate events).

At this point, we are ready for a formal definition of an FSM. An FSM is a quintuple (S, s, E, A, T), where

- S is a finite set of symbols called *states*.

- s is a particular state called the *initial* (or *starting* state).

- E is a finite set of symbols called *events*.

- A is a finite set of symbols called *actions*.

- T is a mapping $T: S \times E \longrightarrow S \times 2^A$ (which maps each {*state, event*} pair into a {*state, set of actions*} pair).

Finally, there exists a versatile language, SDL,[105] standardized by ITU, which is systematically used for detailed protocol description. SDL has the capability to describe FSMs. Because it is used extensively in ITU-T IN Recommendations[106] for the FSM specification, we will briefly introduce three of its graphical constructs that are relevant to the material of this section. Figure 5.3*a* depicts symbols that correspond to a state, an input (an event), and an output (an action). In addition, the actions may be expressed as processing constructs where the usual flowchart symbols are employed. An example of an FSM description using SDL can be found in Fig. 5.3*b*, which presents the FSM of Fig. 5.2.

5.2.5 Call modeling

The call is modeled with the aid of two objects within the CCF. These objects are called the *Originating and Terminating Basic Call State Models (O_BCSM and T_BCSM)*. Note that the term BCSM is—in the parlance of object-oriented programming—"overloaded." In other words, it may mean several things.

First of all, according to clause A.1 of Annex A of Recommendation Q.1204, it is "a high-level Finite State Machine description of CCF activities required to establish communication paths for users [i.e., call

---

[104]Consider, for example, what should really happen when the ringing telephone is answered. Simply moving to the Receiver_Up state, as has been done in our crude design example, does not result in an accurate model because dialing is not what one normally does after one has answered a call. To remedy this problem, one could add a special state for the *ringing* (or *alerting*) activity, in which the event of lifting the receiver causes a transition into the *Talking* state.

[105]See footnote 88 of this chapter.

[106]We have already followed an SDL diagram of Recommendation Q.1213, as related to queuing. Recommendation Q.1214 uses SDL for description of SIBs and BCSM, and Recommendation Q.1218 uses SDL to specify both BCSM and all the rest of IN FE models. (See also footnote 112.)

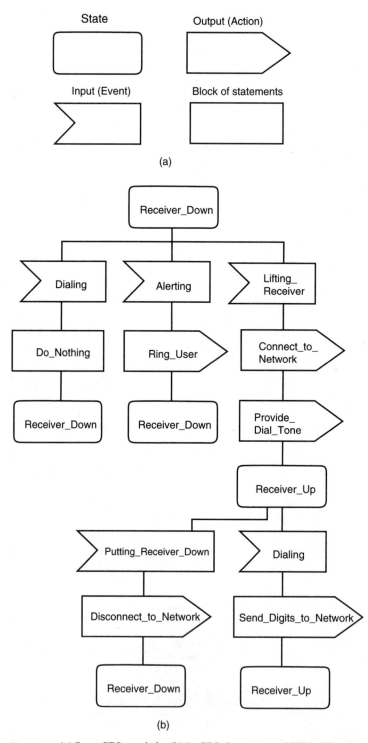

Figure 5.3 (a) Some SDL symbols. (b) An SDL description of FSM of Fig. 5.2.

parties]." Secondly, it applies to the objects that correspond to the originating and terminating parts of the call model (they are respectively called *O_BCSM* and *T_BCSM*). There is, by the way, nothing wrong with using the same term to denote different concepts as long as the context within which the term is used is clear. In the case of BCSM, the objects whose behavior is described by the BCSM FSMs are also called BCSMs. This happens consistently throughout the IN Recommendations. In fact, in the case of the rest of the FEs, the term "state machine" also refers to the objects themselves.

The discussion of POIs and PORs of Chap. 3 has already pointed out the necessity of separating the originating and terminating parts of the call. As far as IN is concerned, we should consider two cases. Figure 5.4*a* depicts the case when the calling and called parties are respectively attached (via CCAF) to O_BCSM and T_BCSM within the same CCF object.

The second case, depicted in Fig. 5.4*b*, deals with the situation when the calling and called parties are attached to different CCF objects. In this case, O_BCSM and T_BCSM have to exchange the information across the network. Note that either object may interact with the SCF (through the SSF) independently of the other, which finally provides a neat insight into how the service logic for originating and terminating services can be designed.

The description of BCSM has its peculiarities, which are as follows:

1. The states are referred to as *Points in Call* (PICs). Graphically, PICs are represented by rectangles.

2. Some transitions are associated with *Detection Points* (DPs), which correspond to combinations of events that may (but do not have to) result in IN processing. If they do, BCSM sends appropriate messages to the SCF and the processing may be suspended at the DPs. Unfortunately, not all BCSM transitions that may take place at a given DP are captured in the model. The present IN model, theoretically, supports transitions to any PIC from any DP, if instructed by the service logic. For this reason, the description of the BCSM that follows is chiefly relevant only to "basic" calls.

Recommendation Q.1204 explicitly calls the BCSM in its Annex A an *example* BCSM. Certain states correspond to what is presently accepted as a strictly switch-based set of functional capabilities, and therefore the BCSM of Recommendation Q.1204 encompasses more than what is presently standardized. (The call models for CS-1 and CS-2 are discussed in Sec. 5.4.) Nevertheless, as an example, it both perfectly illustrates the method used in the standards and provides a clue to the future IN capabilities.

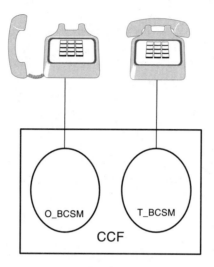

(a)

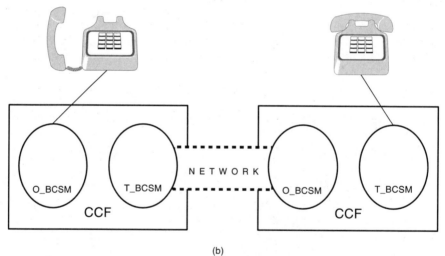

(b)

Figure 5.4 (*a*) Both call parties are attached to the same CCF object. (*b*) Call parties are attached to different CCF objects.

In the description of the BCSM, the Recommendation references, whenever possible, the relation of the BCSM to the external objects with which BCSM interacts.[107] On the originating side, the IN processing of a call in a switch may start when the switch is accessed through

1. A non-ISDN line attached to the switch

2. An ISDN line (attached to an ISDN terminal) or a trunk (attached to a PBX office)

3. A trunk from a non-ISDN switch

4. A trunk from an ISDN switch

Similarly, on the terminating side, the IN processing of a call ultimately results in establishing a connection through

1. A non-ISDN line attached to the switch

2. An ISDN line (attached to an ISDN terminal) or a trunk (attached to a PBX office)

3. A trunk to a non-ISDN switch

4. A trunk to an ISDN switch

In cases 1 and 3, the call-party states pretty much correspond to the position of the receiver (on-hook, off-hook);[108] in case 2, the access *call states* (as well as the rest of the access protocol) are specified in ITU-T Recommendation Q.931;[109] in case 4, the states of the communicating objects and the relevant protocol compose ISUP. This explains how important for IN is the issue of protocol *interworking*. Recommendation Q.1204 is the first to address this issue with specific examples. To this end, the Recommendation carefully considers the associated states and messages of other protocols (particularly, Q.931 and ISUP), although it asserts in clause A.2 of Annex A that its description of the relation of PICs to Q.931 is "not intended to be a detailed formal definition of the relation between the PICs and Q.931 ISDN Call States."

In the remainder of this section, we review O_BCSM, T_BCSM, and the concept of *Call Segments* and *Call Views* as presented in Recommendation Q.1204.

5.2.5.1   Originating BCSM (O_BCSM).   The state machine for the O_BCSM object is depicted in Fig. 5.5 (which is a copy of Fig. A.2/Q.1204). It has 11 PICs and 21 DPs. The PICs and DPs that would

---

[107]These objects and their relation to IN are discussed in Daryani et al. (1992).

[108]The wireless aspects have not been introduced in Recommendation Q.1204 yet.

[109]Stallings (1992) also provides a comprehensive treatment of the subject.

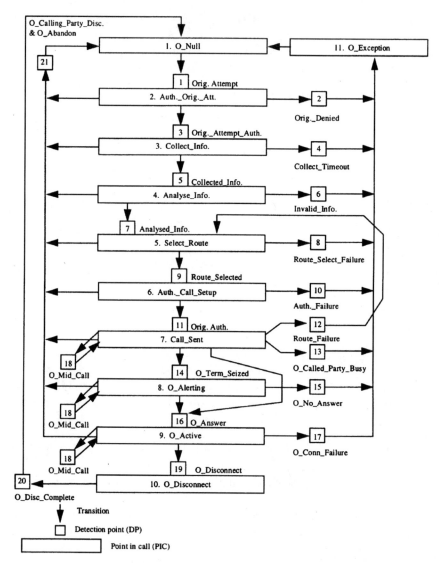

Figure 5.5 Example Originating BCSM. (*After Fig. A.2/Q.1204.*)

otherwise have common names with those of the T_BCSM object are prefaced with the string "O_".[110] Note that O_BCSM has more transitions than DPs, because one DP may be common to several transitions. A good example is DP 21, which corresponds to an event when the calling party disconnects (rather unexpectedly) its telephone.

In what follows, we review O_BCSM, PIC by PIC, briefly describing the DPs associated with each PIC; however, the DPs that are common to several PICs (namely, DPs 18 and 21) are described at the end of this section. Although our goal is to provide material complementary to the text of the Recommendation, the following description is self-contained:[111]

PIC 1: *O_Null*.   At this point, the call does not actually exist. The state is needed to demonstrate the effect of the events that lead to the termination of the call (i.e., all kinds of exceptions or disconnection by a calling party). Otherwise, PIC 2, *Authorize_Origination_ Attempt,* which is reached via DP 1, *Origination_Attempt,* would be the best candidate as the initial state in the life of this object.[112]

PIC 2: *Authorize_Origination_Attempt*.   At this point, the CCF has detected that someone wishes to make a call. Under certain circumstances (e.g., the use of the line is restricted to a certain time of day), such a call may not be placed. The processing required to make the decision is performed in this state. DP 3, *Origination_Attempt_*

---

[110]The names of DPs 13, 15, and 17, when recited in this order, build up a dramatic communications lament: "Oh! Called Party busy...Oh! No answer...Oh! Connection failure..."; the names of DPs 14 and 16 provide a positively optimistic counterpoint: "Oh! Term seized!...Oh! Answer!"; and DP 18 sounds good-naturedly surprised: "Oh! Mid call?". Meanwhile, DP 21 issues somewhat pompous orders: "Oh! Abandon! O, Calling Party, Disconnect!", but the order emitted by DP 19 is laconic: "Oh! Disconnect!"

[111]Please note that the description is also limited to the case of the "basic" call. If a DP is armed, a transition that is not explicitly shown in the standard model may take place. (See the discussion of handling DP 13 in PIC 7.)

[112]It is interesting that the presence of the *Null* state does become an issue when using SDL, because the target unit of an SDL description is a computing *process.* An *O_BCSM* process is created when the switch becomes aware of the originating call; it is destroyed when the call ceases to exist. For this reason, when using SDL, it makes sense to omit the *Null* state. This can be done by introducing to the model a managing process (in practical implementations, usually a part of an operating system) that creates the *O_BCSM* process on a call origination attempt and destroys it when either an exception or disconnection occurs. The model becomes a bit more complex in this case: the managing process ought to handle all external events in the life of the *O_BCSM* processes and decide which events to pass to them and which ones to process by itself. Overall, the SDL descriptions of BCSM in Recommendations Q.1214 and Q.1218 use a different approach, in which DPs are modeled as states. As surprising as this approach may seem at a first glance, it is fully justified by the following consideration. The DPs are places where the external actions of the BCSM take place (i.e., messages are sent to other objects). Until the outcome of such actions is known, the transitions cannot be fully determined (cf. footnote 150), which is why the DPs are modeled as states.

Authorized, results in a transition to PIC 3; DP 2, *Origination_ Denied,* results in a transition to PIC 11.

PIC 3: *Collect_Information.* This is the point at which the dialing string is collected from the calling party, which makes this PIC applicable particularly to local "wireline" switches. If the format of the string is incorrect, or the activity is timed out, DP 4, *Collect_Timeout,*[113] results in a transition to PIC 11. If all is well, DP 5, *Collected_Info,* results in a transition to PIC 4.

PIC 4: *Analyse_Information.* At this point, the complete string is being translated into the routing address. In addition, the *call type* (which denotes whether the call is local, or toll, or international) is determined. The failure of this activity results in a transition via DP 6, Invalid_Information, to PIC 11; its success results in a transition via DP 7, Analysed_Information, to PIC 5.

PIC 5: *Select_Route.* At this point (when the routing address is already known), the actual physical route has to be selected. Note that there may be several physical routes (Members of Technical Staff, 1986) corresponding to a routing address. If the route is unavailable (for example, because of focused network congestion), a transition (via DP 8, Route_Select_Failure) to PIC 11 takes place. Otherwise O_BCSM moves into PIC 6 via DP 9, Route_Selected.

PIC 6: *Authorize_Call_Setup.* Certain service features [for example, *Closed User Group* (CUG)] restrict the types of calls that may originate on a given line (or trunk). This PIC is the point at which relevant restrictions are examined for a given call. If the call is not authorized, the processing of DP 10, Authorization_Failure, is invoked, and the transition to PIC 11 takes place. If the call is authorized, O_BCSM moves, via DP 11, Origination_Authorized, to PIC 7.

PIC 7: *Call_Sent.* At this point, the control over the establishment of the call has been transferred to the *T_BCSM* object, and the *O_BCSM* object is waiting for a signal confirming that either the call has been presented to the called party or that the called party cannot be reached for a particular reason. In the latter case, one of the following actions takes place: (1) if the cause of failure to reach the called party is network congestion, the processing of DP 12, *Route_Failure,* is invoked, and the transition to PIC 5 takes place, thus allowing the switch to select another route, or (2) if the called party is busy, the processing of DP 13, *O_Called_Party_Busy,* is invoked, and the transition to PIC 11 takes place. There are also two

---

[113]The name of this DP does not reflect its use in the case where the string is incorrect. It is important to remember that the string is being examined syntactically, as it may be collected digit by digit, and if its format is wrong, the processing of DP 4 is invoked.

positive alternative outcomes: (1) reception of a message from *T_BCSM* that the called party is being alerted results in the processing of DP 14, *O_Term_Seized,* and a transition to PIC 8, or (2) reception of a message from *T_BCSM* that the called party has answered the call results in the processing of DP 16, *O_Answer,* and a transition to PIC 9 (thus skipping a PIC).

Now, let us consider the handling of the DP 13 beyond the "basic" call. If the DP is armed, the service logic may queue the call (so that it will stay in the present state, that is, PIC 7, until the party is free and then—via the instruction from the SCF—transit to PIC 8) or it may instruct the SSF/CCF to transit to PIC 5 and call another party on a predefined list. Neither transition is *explicitly* specified here but both are possible in IN as long as the service logic requests them in the instruction.

PIC 8: *O_Alerting.*   At this point, *O_BCSM* is waiting for the called party to answer. If that does not happen within the time period known to *T_BCSM,* the latter so informs *O_BCSM,* at which point the processing of DP 15, *O_No_Answer,* is invoked, and a transition to PIC 11 takes place. If the called party answers within the specified time period, T_BCSM sends a respective message to O_BCSM, thus resulting in the processing of DP 16, O_Answer, and a transition to PIC 9.

PIC 9: *O_Active.*   Once this state is reached, the call may become inactive again only when one of the following three events happens: (1) the network connection fails, (2) the *called* party disconnects the call, or (3) the *calling* party disconnects the call. These three events respectively result in processing the following DPs: DP 17, *O_Connection_Failure*; DP 19, *O_Disconnect*; and DP 21, *O_Abandon & O_Calling_Party Disconnect.*[114] The processing of these DPs is accompanied by respective transitions to PICs 11, 10, and 1.

PIC 10: *O_Disconnect.*   As pointed out in the description of PIC 9, the present PIC is reached after the called party had disconnected (and the message containing this information had been issued by *T_BCSM* and subsequently received by *O_BCSM*). After performing all the necessary work (releasing the resources, notifying the calling party of the disconnection of the call, etc.), *O_BCSM* arrives at DP 20, *O_Disconnect_Complete,* which results in a transition to PIC 1.

---

[114]The call is considered *abandoned* by the calling party if it hangs up at any PIC other than PIC 9. If the calling party hangs up while the call is active, its action is considered *disconnection.* This is why the name of the last DP in this sequence, which occurs in all PICs, reflects both actions.

PIC 11: *O_Exception.*   As far as IN is concerned, much clean-up of the network is to be done at this point. A relationship with the SCF, which was created on behalf of the call, has to be destroyed by *both* the SSF/CCF and the SCF (which must, in turn, destroy other relationships that it had created). To achieve that, the SSF/CCF has to issue an appropriate *error* IF to the SCF.[115] In addition, the SSF/CCF should perform the procedures that are necessary to release the lines or trunks involved in supporting the call, which may involve issuing appropriate messages to other network elements. At the end of processing the exception, *O_BCSM* enters PIC 1, the transition having no DP associated with it.

Finally, two DPs are applicable to more than one PIC. These DPs are DP 18, *O_Mid_Call,* and DP 21, *O_Calling_Party_Disconnect & O_Abandon.* The former, also known as a *mid-call trigger,* is applicable to PICs 7, 8, and 9, and reflects a signaling action by a calling party (e.g., a hook flash), indicating that the call has to be temporarily interrupted.[116] The processing of DP 18 results in the transition to the same state. DP 21[117] is common to all PICs except PIC 1 and PIC 11; its processing is always accompanied by a transition to PIC 1.

5.2.5.2   Terminating BCSM (*T_BCSM*).   The state machine for the terminating BCSM object is depicted in Fig. 5.6 (which is a copy of Fig. A.3/Q.1204). The state machine has 8 PICs and 14 DPs. The PICs and DPs that would otherwise have names in common with those of the originating BCSM object are prefaced with the string "T_". As it happens in the case of *O_BCSM* (and for the same reason), *T_BCSM* has more transitions than DPs.[118] Note that the numbering of both the PICs and DPs of *T_BCSM* continues from that of *O_BCSM* rather than starts anew. Thus the first PIC of *T_BCSM* is numbered 12, and its first DP, 22.

---

[115]Recommendation Q.1204 points out a nuance: It may have happened that the SCF had provided *several* instructions to the SSF/CCF before *O_BCSM* entered PIC 11. Some of these instructions could already have been executed, while others remained not acted upon. In the error IF, the SSF/CCF must list these outstanding instructions.

[116]In Credit Card Calling or wireless communications, to name just two examples, it is desirable to allow the customer to restart a new call within the previous context in order to avoid the customer authentication procedure; provision of the mid-call trigger is one way to achieve that. Another obvious use of the mid-call trigger is in support of Conference Calling and other ISDN features (e.g., Call Transfer). Note that the calling party is allowed to use the mid-call capability even before the call enters its active phase (PIC 9); the only prerequisite for supporting this capability is obtaining the authorization for setting up a call.

[117]For the description of DP 21 see footnote 114.

[118]Again (cf. footnote 111), the transitions that are discussed here are limited to the case of the "basic" call. If a DP is armed, a transition that is not explicitly shown in the standard model may take place. (See the discussion of handling DP 25 in PIC 14.)

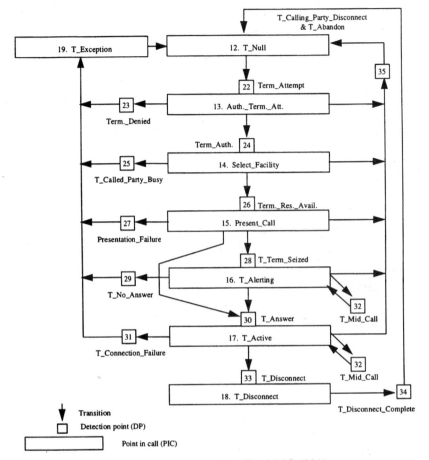

Figure 5.6 Example terminating BCSM. (*After Fig. A.3/Q.1204.*)

In what follows, we review *T_BCSM* in the same way we have reviewed *O_BCSM*, that is PIC by PIC, describing the DPs associated with each PIC. Again an exception is made for the DPs that are common to several PICs (DPs 32 and 35, in this case), which are described at the end of the section.

The PICs of *T_BCSM* are as follows:

PIC 12: *T_Null.*   Everything said about PIC 1, as far as its purpose is concerned, also applies to the present PIC. The call does not exist at this point. A message from *O_BCSM* informing *T_BCSM* about the incoming call results in the processing of DP 22, *Termination_ Attempt,* and a transition to PIC 13.

PIC 13: *Authorize_Termination_Attempt.* At this point, it is verified whether the call is to be passed to the terminating party. While it is the authority of a calling party that is being questioned at PIC 2 (of *O_BCSM*), the intention of the verification at the present PIC is to ascertain that the called party wishes to receive the present type of a call, that its line has no restrictions against this type of call, and that its facilities (bearer capabilities) are compatible with that type of call. If any of these conditions is not met, DP 23, *Termination_ Denied,* is invoked and a transition to PIC 19 takes place; otherwise, the processing of DP 24, *Termination_Authorized,* is invoked, and a transition to PIC 14 takes place.

PIC 14: *Select_Facility.* At this point, the terminating resource (i.e., line or trunk)[119] is being selected. If no resource is available or the called party is busy, the processing of DP 25, *T_Called_Party_Busy,* is invoked, followed by a transition to PIC 19; otherwise (when a resource is available and the called party is not busy) the processing of DP 26, *Terminating_Resource_Available,* is invoked and a transition to PIC 15 takes place.

In the previous section, we made a digression beyond the "basic" call on handling the situation when the called party is busy. Since this situation is first detected at this point, we consider it here as another example of the "nonbasic" call treatment. First of all, such a treatment may be provided by the SSF/CCF itself, which may select another ISDN channel to the called party or even alert it with a "click" or other similar signal. Then the party is to decide whether it wishes to accept the call. In most cases, where such treatment is possible, the *T_BCSM* would simply transit to PIC 19, while the *O_BCSM* would not have to be informed that the called party is busy. Of course, the same and more can be achieved with the IN. Again no such transition is *explicitly* specified here but both are possible as long as the service logic requests them in the instruction.

PIC 15: *Present_Call.* At this point, the call is being presented (via the ISUP ACM message, or Q.931 Alerting message, or simply by ringing in the POTS case). If the replying signaling message indicates that the call cannot be presented, the processing of DP 27, *Presentation_Failure,* is invoked, followed by a transition to PIC 19; if the replying signaling message indicates that the called party is being alerted, the processing of DP 28, *T_Term_Seized,* is invoked,

---

[119]Note that a *T_BCSM* object may reside in an intermediate or originating switch (not only the terminating one), which means that its job in establishing the call is to find a trunk to the next switch on the path to the called party.

followed by a transition to PIC 16; if the call has been answered by the called party, DP 30, *T_Answer*, is processed, and a transition to PIC 17 takes place (thus skipping a PIC).

PIC 16: *T_Alerting.*   At this point, the called party is "alerted" (normally, via ringing).[120] To prevent indefinite holding of the network resources (i.e., trunks and computing resources), which have been acquired by now, this activity must be timed. The expiration of the relevant timer results in the processing of DP 29, *T_No_Answer*, and a transition to PIC 19; if the called party answers, then DP 30, *T_Answer*, is processed, followed by a transition to PIC 17.

PIC 17: *T_Active.*   As the name of the PIC implies, the call enters its active state.[121] Once this state is reached, the call may become inactive only when one of the following three events happens: (1) the network fails the connection, (2) the *called* party disconnects the call, or (3) the *calling* party disconnects the call. Event (1) brings about the processing of DP 31, *T_Connection_Failure,* and a transition to PIC 19; event (2) results in the processing of DP 33, *T_Disconnect,* and a transition to PIC 18; and event (3) results in the processing of DP 35, *T_Calling_Party_Disconnect & T_Abandon,*[122] and a transition to PIC 12.

PIC 18: *T_Disconnect.*   At this point, the *disconnect treatment* associated with called party's having disconnected the call is performed. The end of this activity brings about the processing of DP 34, *T_Disconnect_Complete,* and a transition to PIC 12.

PIC 19: *T_Exception.*   Everything said in the description of PIC 11 of *O_BCSM* applies here.

Similarly to the case of *O_BCSM*, two *T_BCSM* DPs are applicable to more than one PIC. These DPs are DP 32, *T_Mid_Call,* and DP 35, *T_Calling_Party_Disconnect & T_Abandon.* The former is applicable to PICs 16 and 17 and reflects a signaling action by a called party (e.g., a hook flash), indicating that the call has to be temporarily interrupt-

---

[120]The Recommendation also states that at this PIC the *O_BCSM* is being notified about the alerting action; however, technically, this should have happened during the processing of DP 28, since it is a one-time action (cf. the following footnote).

[121]This time the Recommendation correctly states (cf. the previous footnote) that on entering this PIC (as an action performed during the processing of DP 30), *O_BCSM* is being notified of the event.

[122]The event that brings this DP about is the reception of a notification from the *O_BCSM* sent by the latter while processing DP 21, O_Calling_Party_Disconnect & O_Abandon (see footnote 114).

ed.[123] The processing of DP 32 results in the transition to the same state. DP 35 is common to all PICs except PIC 12 and PIC 18; its processing is always accompanied by a transition to PIC 12.

5.2.5.3 Call Segments and Views. Annex C of Recommendation Q.1204 provides a (rather terse) definition of a *Call Segment Model* (CSM) and *Call Views*. In the rest of this section, we will try to formalize and explain this material. The CSM models the part of the connection path contained strictly *inside* a switch as a sequence of three objects, Basic Call Segment, Feature Segment, and Access Segment. The first object in this sequence corresponds to an *O_BCSM* or *T_BCSM*, depending on the role of the connection in the call. The second object corresponds to the service logic process (internal to the switch) invoked on behalf of the call party. The third object is the switch's external interface point, as far this particular connection is concerned.

The subject of *Views* deals with the visibility of the call processing in the switch to the SCF. In the least, the service logic process within the SCF interacts with one Feature Segment within the SSF (and, indirectly, with one Basic Call Segment). This represents the *Local View*.

The *Global View* is the capability of the service logic process to be aware of and, effectively, control all the local views for a given call. This can be achieved by either consolidating the Global View within one service logic process or keeping it distributed. In the latter case, a single service logic process still interacts with only one Feature Segment, but all service logic processes supporting a particular call share the information of the Feature Segments they control.

The extent to which the Views are supported in a given Capability Set is an essential factor in shaping this Capability Set. As the reader may remember from Chap. 3, only Local Views are permitted in CS-1,[124] while CS-2 is progressing toward supporting Global Views.

5.3 Relevant Aspects of Recommendation Q.1211, Introduction to Intelligent Network Capability Set 1

These aspects are as follows: the types of networks that support CS-1 applications, architectural principles, functional relationships, and internetworking.

---

[123]The service requirements in this case are somewhat different from that of *O_BCSM*. The credit card calling requirements, for example, do not apply here. But the mid-call trigger can still be used in support of conference calling, call waiting, and other supplementary ISDN features. In addition, the calling party is allowed to use the mid-call capability only when the call is effectively in its active phase.

[124]As a corollary to single-endedness and single-point-of-control principles of Recommendation Q.1211.

5.3.1   Supporting networks

Recommendation Q.1211 postulates[125] that CS-1 applications are to be supported over the following networks:

1. Public Switched Telephone Network (PSTN)

2. Integrated Services Digital Network (ISDN) (public and private networks)

3. Public Land Mobile Network (PLMN)

5.3.2   Architectural principles

Recommendation Q.1211 repeats the definitions of the IN functional entities and declares[126] the key architectural principles of CS-1 as related to service control, end-user interactions, and service management. These architectural principles delimit the requirements for and responsibilities of the FEs (and, subsequently, the pieces of equipment in which these FEs are implemented).

The five principles (listed in Clause 6.2.1 of the Recommendation) are as follows:

1. The first principle states that the CCF is ultimately responsible "for integrity of, and control of, the local connection at all times." This principle protects the switching products because it acknowledges that, when necessary, the software in the switches may override external commands.

2. The second principle stresses the service independence of the SSF-to-SCF relationship and, based on the nature of this relationship, states that the "CCF and SSF should never contain service logic specific to CS-1-supported services." This principle eliminates any possibility of requesting that manufacturers provide (or otherwise support) the relevant software within a switch, although, of course, it does not prohibit switch-based services (such as VPNs or Freephone).

3. The third principle declares that when either the SCF or the network link that connects it to the SSF/CCF malfunctions, the "SSF/CCF combination should be capable of reverting to a default call completion sequence, with appropriate announcement(s) to the calling and/or called party." This principle protects the interest of network operators, while placing a definitive requirement on all switching products.

4. The fourth principle states that the SSF may not interact with more than one SCF at any given time on behalf of a call party within the context of a particular call. This requirement significantly simpli-

---

[125]In Clause 5.3 of Recommendation Q.1211.

[126]In Clause 6.2 of Recommendation Q.1211.

fies switch implementation, while it appears to have no negative impact on the IN capabilities.

5. The fifth principle permits call handoff capabilities both from SCF to SCF and SSF/CCF to SSF/CCF as long as they do not violate the fourth principle.[127]

In respect to supporting the user interaction with the SRF (i.e., playing the announcements and collecting the digits), Recommendation Q.1211 adjoins[128] three more principles, the first two of which effectively state that the SCF is the *only* FE that is to provide service control instructions to the SRF and SSF/CCF, for which reason "there shall be no direct service control interaction between the SSF and SRF for CS-1-based services." The SSF and SRF are declared "subsidiary to the SCF." (All that means is that the control interface between the SSF/CCF and SRF is *not* standardized.) The third principle states that the SCF should be capable of "suspending processing of a CS-1-based service" on behalf of a call party and later resuming this processing. This otherwise obscure principle alludes to the Service Assist/Hand-off capabilities that are discussed later in this chapter in connection with the Distributed Functional Plane support of the User Interaction SIB.

Finally, two more principles[129] related to service management respectively state that the service creation and management entities (SMF, SCEF, and SMAF) may be used to update the information in the SSF/CCF, SCF, SDF, and SRF as long as these updates do not "interfere with the CS-1-based service invocation or calls that are already in progress," and that the network operators may give service customers the ability to update the "customer-specific information." Recommendation Q.1211 stresses that while "CS-1 must neither exclude nor constrain the capability of service customers to interact directly with customer-specific service management information," the interfaces and mechanisms for doing so are not standardized in CS-1.

5.3.3   Functional relationships

Figure 5.7 (a copy of Fig. 3/Q.1211) contains everything needed to answer a question on whether a functional interface is standardized in CS-1. Each Distributed Functional Plane relationship is assigned a

---

[127]A call handoff is transfer of the support of the call from one FE to another. In CS-1, only one such capability (that involves the transfer of control from one switch to another) is described in detail. The reader will find the description of different aspects of the so-called Service Assist and Service Hand-off capabilities in this chapter and the next one.

[128]In Clause 6.2.2 of Recommendation Q.1211.

[129]Of Clause 6.2.3 of Recommendation Q.1211.

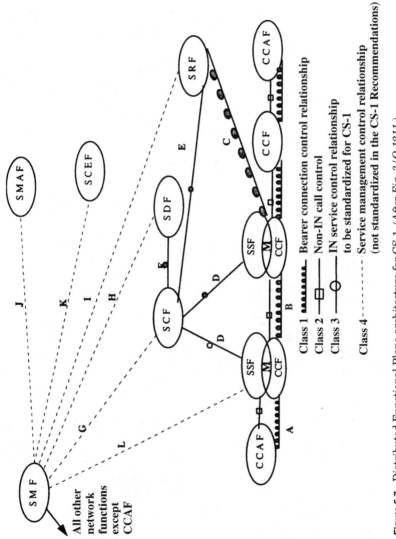

Figure 5.7 Distributed Functional Plane architecture for CS-1. (*After Fig. 3/Q.1211.*)

Class 1 ▬▬▬▬ Bearer connection control relationship
Class 2 ▬▫▬ Non-IN call control
Class 3 ▬○▬ IN service control relationship
          to be standardized for CS-1

Class 4 ------ Service management control relationship
          (not standardized in the CS-1 Recommendations)

capital letter. Only solid lines represent the relationships supported in CS-1. Of these, only the relationships labeled D, E, and F are supported in CS-1. The figure provides further granularity by explicitly depicting the bearer control and non-IN call control relationships.

5.3.4 Internetworking[130]

The relationships discussed in the previous chapter are the *intranetwork* relationships. In other words, the FEs involved are assumed to be located within the same network. Nevertheless, several key CS-1 supported services (e.g., UPT and VPN, to name the two most visible ones) explicitly require the information exchange across the network borders.

For this reason, Recommendation Q.1211 considers five additional functional interfaces for internetworking:

N. SSF-to-SCF

O. SCF-to-SCF

P. SCF-to-SDF

Q. SDF-to-SDF

R. SMF-to-SMF

Of these, only the interface P from an SCF in one network to an SDF in another network is standardized in CS-1.[131] The Recommendation notes that this interface is sufficient for implementation of UPT and VPN because "the translation and validation" can be performed via the information exchange between the SCF in one network and pure database (SDF) in another network.

---

[130]Technically the correct term to use here is "network interworking," and this is the term consistently used in Recommendation Q.1211. Nevertheless, the following circumstances made us decide on using the term "internetworking": (1) By now, it has been dominantly used in both the technical literature, standards discussions, and draft documents and (2) the usage of the term "interworking," on the other hand, is now uniform in references to the mutual influence of one protocol on another (as in INAP/ISUP interworking).

[131]The authors are tempted to include a historical note here. At the time CS-1 was being standardized there seemed to be no existing or planned equipment that would separate the SCF and SDF. (Both could be colocated in a Service Control Point, a Service Node, or an "intelligent switch.") The primary reason for separating the SCF and SDF was internetworking. Indeed, if the SCF and SDF had not been separated, then a general SCP-to-SCP interface would have to be standardized, but this would violate the CS-1 principles and have no solid technological base. (After all, they *became* the principles because they were based on the available technology.) While it was clear how two service logic processes can access the remote data, no such clarity existed with respect to distributed call control. By breaking the SCP into two functions, it was possible to allow SCPs to cooperate on *data* while explicitly prohibiting their cooperation on *call control*. Again, the divide-and-conquer technique worked!

5.4 Recommendation Q.1214, Intelligent Network—Distributed Functional Plane for CS-1

5.4.1 Summary

Recommendation Q.1214 was initially published by ITU-T in March 1993, at which time it had 224 pages. The next version of this Recommendation, CS-1 Refined, was approved in May 1995. The material in this section is based on the final draft text (Thieffry, 1995), which has more than 300 pages.

As its size suggests, Recommendation Q.1214 contains a great deal of material, which includes

1. Clarification of the scope of CS-1 and the associated functional model.

2. FE models of all entities (i.e., SSF/CCF, SCF, SRF, and SDF).

3. The relationships among the FEs and the IFs exchanged across these relationships. This part of the Recommendation also describes the IEs as well as the pre- and postconditions for sending the IFs.

4. Stage 2 description of the GFP SIBs and two other service-independent distributed capabilities, Activity Test and Call Gapping.

In addition, Annex A presents the so-called SSF/CCF relationship scenarios; Annex B supplies the SDL description of BCSM; Appendix I[132] elucidates the DFP aspects that were addressed "for further study" in CS-1 (those aspects that deal with multiparty call handling); and Appendix II provides various charging scenarios.

As far as the CS-1 scope and functional model are concerned, Recommendation Q.1214 is consistent with the material of Recommendation Q.1211, as discussed in the previous section of this chapter.

The FE modeling contains much educational material regarding subdivision of the FEs into modules. Unfortunately, the limitation on the size of this book prevents the authors from discussing these models here; fortunately, the Recommendation's text on that subject is self-contained, and we refer the reader to the text of the Recommendation. Because these models are "static" in that they are not essential to the

---

[132]As we once mentioned, an appendix of an ITU-T Recommendations contains the *informative* text (i.e., the text that is not part of the standard set by Recommendation) versus the *normative* text (i.e., the text that specifies the standard) contained in the Recommendation proper and its annexes.

protocol specification, the omission of this material in the first reading would not cause much of a problem.

The call model, on the other hand, is essential for understanding CS-1, for which reason the authors feel it is important to address it in detail, with the emphasis on DP processing material and the IN-Switching State Model (IN-SSM).[133]

Needless to say, the description of the semantics and vocabulary of the IFs is an important part of the DFP specification,[134] while the description of the service-independent capabilities practically links the IN protocol with the support of services.

The rest of this section discusses the CS-1 call modeling specifics and the service-independent capabilities (with an accent on the IFs used in the distributed processing).[135]

5.4.2   CS-1 call modeling

Although the call model of Recommendation Q.1214, compared to the model of Recommendation Q.1204, supports a much more narrow set of capabilities, it is so rich in detail that the Recommendation even introduces new concepts (which are necessary to sort out the information). These new concepts largely relate to the DP processing,[136] for which reason this section starts with the relevant description, which culminates in building the taxonomy of DPs. After that, the CS-1 BCSM is reviewed, followed by the discussion of the IN Switching State Model

---

[133]The IN-SSM deals with the aspects of connection control that go beyond what is necessary to support the CS-1 capabilities, but the model is part of Recommendation Q.1214, and it defines the terminology and concepts that are further expanded in CS-2.

[134]Here is one reason why it is important. When reaching the Physical Plane of INAP, the builders of the (inverted) Babylonian Tower of the world telecommunications standards started to speak different languages. The evolution of the European implementations ended up with a somewhat different set of particular operations and parameters than what had been specified by Bellcore and implemented by several vendors for a large part of the U.S. market (i.e., Bell Operating Companies). To this end, the ITU-T INAP is composed of both versions of the protocol, each of which may be elected as an option. Both versions, however, can be derived from the same set of DFP's IFs and IEs, which makes one observe that the Q.1214 specification in regard to regional implementations is what Latin language is to, say, French, Italian, Spanish and other Romance languages. As an expert linguist specializing in French needs to know Latin, so a telecommunications engineer who is working on IN is advised to learn the IFs and IEs of Q.1214. (On an encouraging side, the required effort is significantly smaller in this case than what learning Latin demands.)

[135]The material of the appendices to the Recommendation is not discussed in this book; however, the Call Party Handling (CPH) material of the emerging CS-2 Recommendation Q.1224, which is introduced later in this chapter, supersedes the contents of Appendix I.

[136]Which is also responsible for most of the material added to Recommendation Q.1214 during the CS-1 refinement process.

(IN-SSM),[137] which sheds more light on the subject of Call Segments and Views.

5.4.2.1 DP Processing. Each DP may be either *armed* or *not armed*. As far as IN is concerned, being armed is the first essential prerequisite for being active, for only when a DP is armed is the external service logic (within the SCF) informed that the DP is encountered.[138] A DP may be armed either *statically* (from the SMF, as the result of the service feature provisioning) or *dynamically* (by the SCF). In the former case, the DP is staying armed until the SMF disarms it, which means that it is going to be armed for about as long as the service that needs it is to be offered; in the latter case, the DP is staying armed for no longer than the duration of a particular SCF-to-SSF relationship. (This explains why the words *statically* and *dynamically* are used.) A statically armed DP is called a *Trigger Detection Point* (TDP); a dynamically armed DP is called an *Event Detection Point* (EDP).

But even if a DP is armed, it has to satisfy certain DP criteria (which are associated with that DP) in order for the SSF/CCF to emit an IF to the SCF. In the case of an EDP, such criteria are specified in the IF (from the SCF) that is arming the EDP.[139] In the case of the TDP, the criteria may apply to

- An individual subscriber's line or a trunk (*line / trunk-based criteria*)
- A group of lines or trunks (*group-based criteria*)
- The whole switching office (*office-based criteria*)

The 15 CS-1 TDP Criteria types are as follows:

1. *Trigger Assigned* (conditional or unconditional). If it is unconditional, no other criteria are checked; if it is conditional, the associated criteria have to be satisfied.

2. *Class of Service.* This is a code that specifies certain attributes of a line (e.g., payphone) or trunk group (e.g., type of signaling).

3. *Specific B-channel Identifier.*

4. *Specific Digit Strings.*[140]

---

[137]This model is first introduced in Recommendation Q.1214.

[138]But even then, as the reader will soon find out, encountering an armed DP may not result in invoking external service logic. Thus, being armed is a *necessary* but not *sufficient* condition for starting distributed processing of a DP.

[139]For example, the *Request Report BCSM Event* IF specifies a list of events that are to be reported whenever one of them occurs for a given call.

[140]Interestingly, the specified string may be empty (i.e., its size may be specified as zero) to indicate an off-hook trigger.

5. *Feature Codes* (e.g., *69, which is used in North America to call back the previous caller).

6. *Prefixes.*

7. *Access Codes* (for customized numbering plans).

8. *Specific Abbreviated Strings for Customized Numbering Plan.*

9. *Specific Calling Party Number Strings.*

10. *Specific Called Party Number Strings.*

11. *Nature of Address.* An indicator of whether the called party number is local, national, or international, etc.

12. *Bearer Capability* (e.g., DTMF or rotary).

13. *Feature activation/indication* (e.g., feature activation on an ISDN line or a hook-flash on a non-ISDN line).

14. *Facility Information* (as specified in a signaling message).

15. *Cause* (as specified in a signaling message).

As far as the call processing is concerned, either of the two actions may be requested of the SSF/CCF when a DP is encountered:

1. The request for instructions (to the SCF) is issued and the call processing is suspended until the response is received.

2. The call processing continues and notification of the event is sent (the SCF).

Accordingly, two types, *R* (for request) and *N* (for notification), are defined for the DPs. Both EDPs and TDPs must be assigned either type, but it is important to remember that the assignment takes place *only* at the time a particular DP is armed. If a TDP is armed as *R,* it is denoted TDP_R; if it is armed as *N,* it is denoted TDP_N. Similarly, an EDP may be denoted EDP_R or EDP_N, depending on its type. It is important to remember that in order to arm or change a TDP, switches have to be reprovisioned (which is, again, outside of the scope of the IN protocol,[141] at least for CS-1), but arming or changing an EDP is comparatively easy: only one message from the SCF to SSF/CCF is required. Notably, Wireless Intelligent Network (WIN) eliminates this restriction (Cellular Telecommunications Industry Association, 1996): in WIN, provisioning of TDPs is performed dynamically.

---

[141]And, consequently, cannot be supported by service logic.

Figure 5.8 (a copy of Fig. 4-[E] 10/Q.1214) depicts different outcomes of the DP processing in view of the elements of the switching call model. The latter has two new entities: the Basic Call Manager (BCM) and the Feature Interaction Manager (FIM). The initial processing of a BCSM DP is performed by the BCM, which spans both the CCF and SSF. If the DP is not armed, the BCM resumes the processing of the call. Otherwise, the BCM determines the type of the DP and passes its processing to the FIM. The FIM determines whether a message should be sent to the service logic (within the SCF), and, if so, formulates that message. The FIM is also the first element to receive messages from

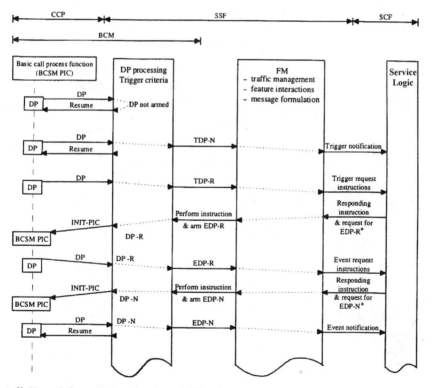

*In this example, the responding instruction and request for EDP are shown together. These are independent information flows and may not be sent together in all cases.

DP	Detection point
TDP	Trigger detection point
EDP	Event detection point
R/N	Request/notification
PIC	Point in call

Figure 5.8 DP processing. (*After Fig. 4-[E] 10/Q.1214.*)

the SCF. Certain call-unrelated messages (such as Call Gap[142]) are not passed any farther but are processed within the FIM.

There are several reasons why a message may not be sent to the SCF. One is that Call Gapping[143] has been applied. The other is that there are other instances of service logic execution, which are incompatible with the request for a particular instruction. Recommendation Q.1214 provides (in Clause 4.2.4.1) specific rules for managing the service logic instance interaction and determining the compatibility of service features.[144]

One more complexity which the SSF/CCF[145] must deal with is caused by the fact that a DP may be armed as both EDP and TDP, and the DP may be of both types. For this reason the FIM should be able to choose the precedence in processing all the hats a particular DP is wearing. To this end, Recommendation Q.1214 postulates that

1. Notifications are to be processed before requests.

2. Events are to be processed before triggers.

3. If a DP is armed as both EDP_R and TDP_R, the latter may be processed only if the control[146] relationship between the SSF and CCF has been terminated as the result of processing EDP_R.

These rules, augmented by the requirement that a specific trigger condition may result in an invocation of only one service logic process, enforce the single-point-of-control principle.

5.4.2.2 CS-1 BCSM. Both the originating (Fig. 5.9[147]) and terminating (Fig. 5.10[148]) BCSMs have fewer PICs than their example counterparts of Recommendation Q.1204. On the other hand, here the capabilities of

---

[142]If the SCF is overloaded, it issues the Call Gap IF to the SSF/CCF, which instructs the switch to reduce the rate of the messages sent to the SCF. Gapping can also be applied "manually" via the SMF. Call Gapping is a network management capability, which is further discussed in Sec. 5.4.3.3.2. A similar capability that may influence DP processing is Service Filtering (see the definition of the Limit SIB of the previous chapter and its description in the present one).

[143]This capability is discussed in Sec. 5.4.3.3.2.

[144]It is unfortunate that no more time can be spent on this subject in this book. We do recommend that the reader with a need to know more about the subject consult the text of the Recommendation. It should be also noted that the subject itself requires a separate book (which would be much larger than the present one).

[145]More precisely, the BCM.

[146]The Recommendation defines the *control* relationship as the one that has more than one armed EDP_R. If the relationship has no armed EDP_Rs, but at least one armed EDP_N, it is called a *monitor* relationship.

[147]This figure is a copy of Fig. 4-3/Q.1214.

[148]This figure is a copy of Fig. 4-4/Q.1214.

the model are described in much more detail. In particular, the description of the model includes the following information:

- Data that are invariant of PICs (i.e., available at all PICs) as well as the data particular to and available at each individual PIC[149]

- Trigger criteria that may be assigned to a particular DP

- Additional (to those depicted in the figures) transitions[150] that may take place at a particular DP

- Non-IN BCSM transitions[151]

- Information on both the *O_BCSM-* and *T_BCSM*-to–end-user access signaling exchange

- Information on signal exchange between the *O_BCSM* and *T_BCSM* objects (located in the two segments of the same SSF/CCF), which correlates the states and transitions of both state machines by demonstrating how a (non-IN) action of one results in an event of another

In what follows, *O_BCSM* and *T_BCSM* for CS-1 are briefly discussed. In the discussion, we concentrate on (1) the differences between the CS-1 BCSM and the example BCSM counterpart of Recommendation Q.1204 as far as PICs and DPs are concerned, (2) the assignment of the trigger criteria to DPs, and (3) additional state transitions (both non-IN BCSM transitions and CS-1 IN transitions beyond a basic call). Note that the midcall trigger,[152] although part of CS-1, is given no detailed treatment by the Recommendation.

---

[149]The data specification has been carefully revised as the result of the CS-1 refinement process. Certain data, which include, for example, *Terminal Type* (i.e., the type of the end user's telephone equipment), *Service Address Information*, etc., are specific to all PICs, while others, like *Bearer Capability, Calling Party Number*, etc., are not. Overall, these data compose an essential part of the distributed data model necessary for the definition of the protocol, since the protocol messages (IFs) are to carry (in its IEs) some of these data from one FE to another. The value of every single datum is set via either service management or CS-1 IFs.

[150]These transitions are referred to as the "IN Transitions Beyond a Basic Call for IN CS-1," because they may occur only as the result of the instruction issued to the SSF/CCF by the SCF to jump to a certain PIC called a "resume point." To properly model this capability in the DFP, one has to add all the transitions associated with the DP so as to end up with what is called a *nondeterministic automaton,* because the precise transition (out of the several with which a DP is associated) that should take place cannot be determined at the level of the FSM description. In this case, it is the SDL diagrams that provide the needed information. Note, however, that the SDL description of the BCSM expands the model significantly. As we had noted before (see footnote 112), in the expanded model the DPs are *states.* In this case, the instruction from the SCF to go to a specific PIC is processed at the same DP at which the original request for instruction was issued.

[151]These transitions take place on the CCF events (such as finding out that a certain line or trunk is busy) that are neither caused nor require IN processing in CS-1.

[152]DP 8, *O_Mid_Call.* Note, by the way, that in CS-1 it may be processed only in PIC 5, *O_Active.*

5.4.2.2.1 *O_BCSM*. *O_BCSM* specifies six PICs and ten DPs. Comparing the present *O_BCSM* to the example *O_BCSM* (Fig. 5.5), we observe that PICs 1 and 2 of the latter are consolidated into one PIC (a new PIC 1, called *O_Null & Authorize Origination Attempt*), and PICs 5 through 8 are consolidated into a new PIC 4, called *Routing & Alerting*.

One non-IN event,[153] *Route_Busy,* is considered in *O_BCSM*. The event is caused by either a corresponding indication from *T_BCSM* (if this is a local switch) or a reception of the call rejected message (indicating that the selected route is busy) from another switch. Depending on the switch application, this event may result in either a transition to PIC 2 (in order to select another route) or exception processing.

---

[153]This event was added as the result of the CS-1 refinement process.

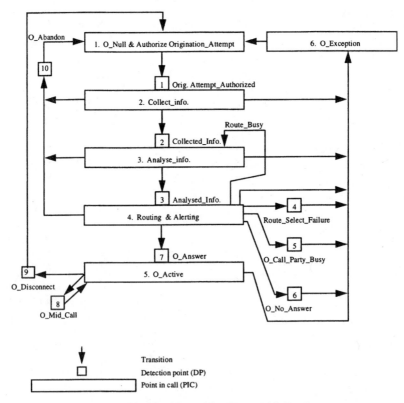

Figure 5.9 Originating BCSM for CS-1. (*After Fig. 4-3 / Q.1214.*)

The IN transitions beyond a basic call are associated with DPs 1, 2, 3, 4, 5, 6, and 9. Each of these seven DPs may cause a transition to PICs 2, 3, and 4.[154]

Finally, as far as the applicability of the CS-1 TDP criteria to DPs is concerned, Recommendation Q.1214 grades it as applicable, optional (i.e., applicable under certain circumstances[155]), or not applicable. The following list specifies the grade of applicability of each criterion to each of the DPs of O_BCSM:[156]

1 The *Trigger Assigned* criterion is applicable to DPs 1 through 10.

2. The *Class of Service* criterion is applicable to DP 1 and optional at DPs 2 through 10.

3. The *Specific B-channel Identifier* criterion is optional at DPs 1 through 10.

4. The *Specific Digit Strings* criterion is not applicable to DP 1, but applicable to DPs 2 and 3, and optional at DPs 4 through 10.

5. The *Feature Codes* criterion is not applicable to DP 1, but applicable to DPs 2 and 3, and optional at DPs 4 through 10.

6. The *Prefixes* criterion is not applicable to DP 1, but applicable to DPs 2 and 3, and optional at DPs 4 through 10.

7. The *Access Codes* criterion is not applicable to DP 1, but applicable to DPs 2 and 3, and optional at DPs 4 through 10.

8. The *Specific Abbreviated Strings for Customized Numbering Plan* criterion is not applicable to DPs 1 and 2, but applicable to DP 3, and optional at DPs 4 through 10.

9. The *Specific Calling Party Number Strings* criterion is applicable to DP 1 and optional at DPs 2 through 10.

10. The *Specific Called Party Number Strings* criterion is not applicable to DP 1, but applicable to DPs 2 and 3, and optional at DPs 4 through 10.

11. The *Nature of Address* criterion is not applicable to DPs 1 and 2, but applicable to DP 3, and optional at DPs 4 through 10.

---

[154]The reader has probably noticed that this set actually includes three basic call transitions (namely, a transition to PIC 2 at DP 1, a transition to PIC 3 at DP 2, and a transition to PIC 4 at DP 3). We hope that this laxity is compensated for by the simplicity of the mnemonic.

[155]These circumstances (i.e., presence of certain information at a particular DP), which are sometimes implementation dependent, are specified in Clause 4.2.2.5 of Recommendation Q.1214.

[156]Note that no DP criterion is applicable to DP 11.

12. The *Bearer Capability* criterion is applicable to DP 1 and optional at DPs 2 through 10.

13. The *Feature Activation/Indication* criterion is not applicable to DPs 1 and 2, but applicable to DPs 3 through 10.

14. The *Facility Information* criterion is not applicable to DPs 1, 2, 4, 5, 6, 9, and 10, but applicable to DPs 3, 8, and 9.

15. The *Cause* criterion is not applicable to DPs 1, 2, 3, 6, 7, and 8, but applicable to DPs 4, 5, 9, and 10.

5.4.2.2.2 *T_BCSM*. *T_BCSM* specifies five PICs (numbered 7 through 11) and seven DPs (numbered 12 through 18). Comparing the present *T_BCSM* to the example *T_BCSM* (Fig. 5.6), we observe that PICs 12 and 13 of the latter are consolidated into one PIC (a new PIC 7, called *T_Null & Authorize_Termination_Attempt*), and PICs 14 and 15 are consolidated into a new PIC 8, called *Select_Facility & Present_Call*.

Note that a transition from PIC 8 to PIC 9 has no DP associated with it. The event that causes this transition (i.e., the start of the alerting process) is the only strictly non-IN event visible to *T_BCSM*.

The IN transitions beyond a basic call are associated with DPs 13 and 14. Each of these two DPs may cause a transition to PIC 8.

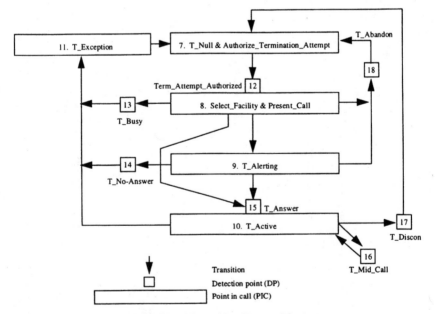

Figure 5.10 Terminating BCSM for CS-1. (*After Fig. 4-4/Q.1214.*)

The following list specifies the grade of applicability of each DP criterion to each of the DPs of *T_BCSM*.

1. The *Trigger Assigned* criterion is applicable to all DPs.

2. The *Class of Service* criterion is applicable to DP 12 and optional at DPs 13 through 18.

3. The *Specific B-channel Identifier* criterion is not applicable at DPs 12 and 13, but optional at DPs 15 through 18.

4. The *Specific Digit Strings* criterion is not applicable to any DP.

5. The *Feature Codes* criterion is not applicable to any DP.

6. The *Prefixes* criterion is not applicable to any DP.

7. The *Access Codes* criterion is not applicable to any DP.

8. The *Specific Abbreviated Strings for Customized Numbering Plan* criterion is not applicable to any DP.

9. The *Specific Calling Party Number Strings* criterion is not applicable to DP 12, but is applicable to DPs 13 through 18.

10. The *Specific Called Party Number Strings* criterion is not applicable to any DP.

11. The *Nature of Address* criterion is not applicable to any DP.

12. The *Bearer Capability* criterion is applicable to DP 1 and optional at DPs 2 through 10.

13. The *Feature Activation/Indication* criterion is not applicable to DPs 12 and 13, but is applicable to DPs 14 through 18.

14. The *Facility Information* criterion is not applicable to DPs 12 through 14, but is applicable to DPs 17 and 18.

15. The *Cause* criterion is not applicable to DPs 12, 14, 15 and 16, but is applicable to DPs 13, 17, and 18.

5.4.2.3   IN Switching State Model (IN-SSM).   The IN-SSM is defined[157] as a means of providing the "finite state machine description of SSF/CCF IN call/connection processing in terms of call/connection states." Although the CS-1 IN-SSM description is not finite-state-machine-based, it introduces several important concepts.

The IN-SSM is a class of objects that corresponds to the SCF *view* of call and connection processing within the SSF/CCF. The call segments defined in Recommendation Q.1204 (and reviewed in Sec. 5.2.5.3 of this

---

[157]In Clause 4.2.3 of Recommendation Q.1204.

chapter) are expanded to include the new types of objects called *Legs* and *Connection Points* as well as the BCSM objects (i.e., *O_BCSM* or *T_BCSM*) associated with them. A leg represents a connection with an "addressable entity," that is, an end user, the SRF, or another SSF/CCF; a connection point is the object that associates two legs so that the information entering the SSF/CCF through one leg is carried out (via that object) on the other leg and vice versa. A connection point may actually associate more than two legs (to make, for example, a conference call).

Any leg may be either *active* or *passive*. In CS-1, only a leg that represents an access interface may be active. Only one leg may be active in an IN-SSM.

Essentially, the IN-SSM issues are primarily of a theoretical interest as far as CS-1 is concerned. The leg manipulations become a serious issue only when a single-endedness restriction is lifted. Recommendation Q.1214 discusses the IN-SSM in light of CS-1 mainly to introduce the terminology that is later used to designate some IEs. On several occasions, the Recommendation admits that the applicability of certain IN-SSM aspects to CS-1 is left for further study.

To conclude this section, we cite a nontrivial example of a three-party call that theoretically may take place in CS-1 if a capability called *segment association* is supported. (The example is depicted in Fig. 5.11, which is a copy of Fig. 4-13/Q.1214.) Two call segments are called *asso-*

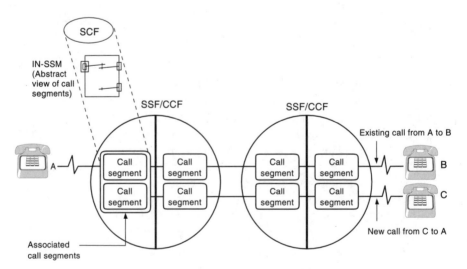

Figure 5.11   *(After Fig. 4-13 / Q.1214.)*

ciated if they can be treated by the SSF/CCF as a single call segment.[158] In the present example, the two segments are extended to user A, who is presently participating in a call with user B. User A may also receive a call from user C (or start a call to user C) via another segment associated with the already active one. The IN-SSM (in the upper left corner of the figure) represents the SCF view of this situation.

5.4.3   Realization of service-independent capabilities in the DFP

The description in the present section parallels the one used in the CS-1 SIB review of the previous chapter. We start with the BCP, and continue with all CS-1 SIBs, introducing the new information flows as they are needed. Except for a few instances when it is absolutely necessary for understanding the crux of the abstract protocol, we do not discuss the IEs. Clause 6 of Recommendation Q.1214 is the repository of the IF vocabulary where the reader will find all the information that is available.

For what are usually referred to as "historical reasons," Recommendation Q.1214 adopted a set of literals to denote the FE-to-FE relationships that differs from the one selected in Recommendation Q.1211 (see Sec. 5.3.3). Table 5.1, which brings both together, should aid the reader.

5.4.3.1   BCP.   While the IN GFP views the BCP as a monolithic process, the Distributed Functional Plane considers it a set of SSF/CCF objects, with which the SCF exchanges information. The relevant set of originating and terminating BCSM objects defines the state of BCP.

---

[158]In other words, there is a mechanism by which the SSF/CCF can be told to associate two given call segments. Recommendation Q.1214 notes that in CS-1 the segments must belong to the same end user. A user may have a telephone terminal with several lines; but even if a terminal has only one physical line, several "virtual" lines can be maintained by the switch, so that the user can alternate these lines in some way (e.g., by means of a flash button).

TABLE 5.1   The FE Relationship Nomenclatures

Relationship	Q.1214 Nomenclature	Q.1211 Nomenclature
CCAF-to-SSF/CCF	r1	A
SSF/CCF-to-SSF/CCF	r2	B
SSF/CCF-to-SCF	r3	D
SCF-to-SDF	r4	F (or P, for internetworking)
SSF/CCF-to-SRF	r5	C
SCF-to-SRF	r6	E

Recommendation Q.1214 defines the BCP capabilities (which are *call setup, call party handling,*[159] *call initiation, call clearing,* and *event reporting*) and notes that these capabilities can be invoked by *either* the SSF/CCF *or* the SCF. This is an important point, which—as our experience has shown—is still often overlooked in interpreting the standard. Indeed, in CS-1 the SCF *may* start IN processing by sending to the SSF/CCF the *Initiate Call Attempt IF*. This is the only example of an "unprovoked" instruction from service control to a switch.

In the rest of this section, only those IFs that are immediately relevant to the BCP capabilities are reviewed. In addition, we review the relation of the POIs and PORs to the BCSM.

On the SSF/CCF side, the IN processing is initiated (i.e., an information flow is sent to the SCF) when a trigger is detected. The SCF receives the request with the information about the call. This information, however, is particular to the DP at which a trigger was encountered. For example, if this DP is *Origination_Attempt_Authorized,* the digits dialed by the calling party may be relevant to the IN processing of the DP, but if the DP is *T_Disconnect,* it is clear that the digits are not needed.[160]

There are two alternative approaches to requesting the SCF processing: one approach is called DP-*generic* and the other, DP-*specific.*

With the DP-generic approach, only one IF, *Initial* DP, is used. The set of information elements that the IF carries depends on the DP encountered.

With the DP-specific approach, there is a separate IF for each DP, but the set of IEs associated with each such IF is limited precisely to what is applicable to the affiliated DP. The DP-specific family of initial IFs is as follows:

Origination Attempt Authorized

Collected Information

Analyzed Information

Route Select Failure

O_Called_Party_Busy

O_No_Answer

O_Answer,

O_MidCall

O_Disconnect

Term Attempt Authorized

T_Called_Party_Busy

---

[159]In CS-1, however, this capability was left for further study.

[160]Indeed, the *Dialed Digits* IE is used in the former case but omitted in the latter one.

T_No_Answer

T_Answer

T_MidCall

T_Disconnect.[161]

Once the initial DP has been sent and the SSF/CCF-to-SCF relation, *r3,* has been established,[162] the SCF may issue instructions to the SSF/CCF or request reports (by arming certain DPs as EDP_R via Request Report BCSM Event IF), or do both in an unlimited sequence, the potentially recursive character of which will be elucidated shortly.

To finish with the BCP-related IFs sent from the SSF/CCF, we observe that they are parallel to the initiating IFs. Again, they can be either DP-generic or DP-specific. In the former case, there is only one IF, Event Report BCSM; in the latter case, the DP-specific family of IFs includes the same IFs as the initial ones listed above.[163] Note though that the semantics of these IFs are different from the initial ones in two ways. First, they are sent in response to a request from the SCF. Secondly, each of these IFs may be—if the DP has been armed as an EDP-N—a notification (which requires no suspension of call processing and waiting for an instruction) or—if the DP has been armed as EDP-R—a request.

The rest of the BCP-related IFs are sent from the SCF to the SSF/CCF.

The *Connect* IF completes the call setup by instructing the SSF/CCF to terminate a call at a specific destination. Antithetically, the *Release Call* IF requests that the SSF/CCF destroy the call no matter what phase it is in.

Finally, there are five IFs issued by the SCF that influence the continuation of the call processing from the DP at which the SSF/CCF had suspended it prior to issuing a request for instructions. The following four IFs[164] instruct the SSF/CCF to resume the call at their namesake PICs:

1. *Collect Information* IF instructs the SSF/CCF to prompt the calling party for the destination address.[165] It may be issued when the BCP is suspended at any DP preceding the active phase (i.e., when it is suspended at any DP numbered 1 through 6).

---

[161]Thus the DP-specific family of IFs has one IF for each DP except for two DPs, *O_Abandon* and *T_Abandon*. The absence of these two seems to be an oversight, which will be fixed in CS-2.

[162]Note that this is the only relation considered as far as BCP is concerned.

[163]See footnote 161.

[164]These IFs compose the *Proceed with Call Processing* family.

[165]This capability is required by VPN service.

2. *Analyze Information* IF instructs the SSF/CCF to resume the BCP at PIC 3. As in the previous case, it may be issued when the BCP is at any DP numbered 1 through 6.

3. *Select Route* IF instructs the SSF/CCF to resume the BCP at PIC 4. Its precondition specifies that the BCP may be at any DP numbered 1 through 6 as well as at DP 9 (*O_Disconnect*). In the latter case, however, it must be the called party who had disconnected the call.[166]

4. *Select Facility* IF instructs the SSF/CCF to resume the BCP at PIC 8.[167] It may be issued when the BCP is suspended at any DP numbered 12 through 14.

The fifth IF related to the BCP processing is *Continue*. It simply instructs the SSF/CCF to continue processing at the same point the BCP was suspended and with the information that is presently available. The usefulness of this IF is that it spares sending the redundant (i.e., already available at the switch) data, which would have been the case if any other IF had been issued in response to the request from the SSF/CCF.

Now we shall revisit the GFP's POIs and PORs, which we discussed at some length in the previous chapter. Fortunately, no elaborate modeling is required in the Distributed Functional Plane to elucidate this concept. The model is realized as follows:

- The POIs correspond to DPs (more precisely, to the DPs of the TDP-R type), and the PORs correspond to PICs.

- When the processing is suspended at a DP (POI), the SSF/CCF formulates a message and issues a request for instructions to the SCF. This is done by the Feature Interaction Manager (FIM) (recall Fig. 5.8 and the accompanying discussion). It is then the FIM—not the BCP—that issues what amounts to the remote procedure call.

- When the FIM receives the response,[168] it—among other things—instructs the Basic Call Manager (BCM) to resume the call processing at a specific point.

The mapping of POIs and PORs presented in Clause 5.5.1 of Recommendation Q.1214 is as follows:

1. *Call Originated* POI corresponds to DP 1, *Origination_Attempt_Authorized.*

---

[166]This is one nontrivial example of a powerful capability (to place a sequence of calls automatically). The capability is supported by a specifically IN (versus BCP) transition.

[167]Of the four members of the family, this IF is the only one that affects the *T_BCSM*.

[168]And, if the remote procedure call model is employed, that happens at exactly the same place in the processing where it was issued.

2. *Address Collected* POI corresponds to DP 2, *Collected_ Information.*

3. *Address Analyzed* POI corresponds to DP 3, *Analyzed_ Information.*

4. *Prepared to Complete Call* POI corresponds to DP 12, *Termination_Attempt_Authorized.*

5. *Busy* POI corresponds to DP 4, *Route_Select_Failure,* DP 5, *O_Called_Party_Busy,* and DP 13, *T_Busy.*

6. *No Answer* POI corresponds to DP 6, *O_No_Answer,* and DP14, *T_No_Answer.*

7. *Call Acceptance* POI corresponds to DP 7, *O_Answer,* and DP 15, *T_Answer.*

8. *Active State* POI corresponds to DP 8, *O_Mid_Call* and DP 16, *T_Mid_Call.*

9. *End of Call* POI corresponds to DP 10, *O_Abandon,* DP 18, *T_Abandon,* DP 9, *O_Disconnect,* and DP 17, *T_Disconnect.*

As far as the PORs are concerned, they are mapped to PICs as follows:

1. *Continue with Existing Data* POR corresponds to the PIC reached via the transition associated with the DP at which the BCP was suspended.[169]

2. *Proceed with New Data* POR corresponds either to the PIC reached via the transition associated with the DP at which the BCP was suspended,[170] or one of the following four PICs: PIC 2, PIC 3, PIC 4, or PIC 8.[171]

3. *Handle as Transit* POR corresponds to either PIC 3 or PIC 4.

4. *Clear Call* POR corresponds to either PIC 1 or PIC 7.

5. *Enable Call Party Handling* POR, as far as the CS-1 (in which Call Party Handling is not supported) is concerned, corresponds to the PIC reached via the transition associated with the DP at which the BCP was suspended.

6. *Initiate Call* POR corresponds to either PIC 3 or PIC 4.[172]

---

[169]Cf. the description of the *Continue* IF above.

[170]In this case, the SCF response has provided the new data related to the call.

[171]Cf. the description of the *Collect Information, Analyze Information, Select Route,* and *Select Facility* IFs above.

[172]When the SSF/CCF receives the Initiate Call Attempt IF from the SCF, a new instance of the BCSM is created, and its state is determined to be either PIC 3 or PIC 4 depending on the information carried by this IF (as is the case with most IFs, the majority of the IEs here are optional).

In conclusion, we note that Recommendation Q.1214 does not explicitly map the Service Support Data (SSD) and Call Instance Data (CID) into the information model of the DFP.

5.4.3.2 CS-1 SIBs. The Stage 2 SIB description in Clause 5.2 of Recommendation Q.1214 systematically adheres to a template, which, for each SIB, contains the following items (in that order):

1. A short description of the SIB

2. A graphical depiction of the information exchange[173] among the FEs over the involved relationships with the references to the FEAs performed at each FE[174]

3. The list of IEs carried by each IF involved

4. The SDL description of each FE, which links the reception of an IF with the corresponding FEA

5. The prose description of all FEAs

The purpose of the following review is to continue the introduction of the CS-1 IFs while demonstrating how the SIBs can be implemented in principle.

5.4.3.2.1 Algorithm SIB. This SIB is realized entirely within the SCF.

5.4.3.2.2 Authenticate SIB. This SIB is realized through the SCF-to-SDF relationship *r4*. The SCF issues the *Authenticate* IF to the SDF, and the SDF responds with the *Authenticate Result* IF to the SCF.

5.4.3.2.3 Charge SIB. In non-IN networks charging is the function of the BCP. The flexibility of feature introduction supported in IN demands that the service logic participates in the charging process to ensure that the charging is consistent with the service requirements. To this end, Recommendation Q.1214 specifies four types of charging scenarios, which "may be used in any combination appropriate for a given service and given network." All four scenarios realize the charging SIB via processing at the SSF/CCF and SCF and the information

---

[173]In general, such an exchange reflects a "sunny day" scenario, which does not take into account possible errors, exceptions, or their consequences. Some (but not all) of the latter are captured in the SDL diagrams that follow. (The detailed protocol procedures are specified in Recommendation Q.1218, although there they are not aligned with SIBs.) Nevertheless, the absence of the "rainy day" material is not necessarily a drawback because the ultimate goal of Recommendation Q.1214 was not to implement SIBs but to demonstrate how the abstract IN protocol *can* be used to do that. To this end, in the authors' opinion, the Recommendation has succeeded in providing a clear and convincing picture.

[174]Note, however, that this item is not applicable to the description of those SIBs, whose execution is confined to the SCF.

exchange over the SSF/CCF-to-SCF relationship *r3*. In the first two scenarios, the SCF is not involved in maintaining charging data or charging events; in the last two, it keeps track of both.

In the first scenario, the SSF/CCF is responsible for charging. The SCF issues the *Furnish Charging Information* IF, which carries the billing characteristics for a particular call, to the SSF/CCF. Upon reception of this IF, the SSF/CCF generates a billing record.

In the second scenario, the "network charging functions"[175] are responsible for the maintenance of charging information, which they receive from the SSF/CCF. The latter is prompted to send this information by the SCF. Thus the SCF issues the *Send Charging Information* IF, which carries the billing characteristics for a particular call, to the SSF/CCF. Upon reception of this IF, the SSF/CCF sends the charging information to an appropriate network element.

In the third scenario, the SCF monitors charging events[176] (which are detected at the SSF/CCF). The SCF issues the *Request Notification Charging Event* IF, which specifies an event associated with a particular call, to the SSF/CCF. Upon receiving the IF, the SSF/CCF starts monitoring for the event, and when it is detected, it sends the *Event Notification Charging* IF to the SCF.

In the fourth scenario, the SCF is becoming involved with both the specifics of charging procedures performed by the SSF/CCF[177] and keeping track of charging events. The SCF issues the *Apply Charging* IF, which carries the charging characteristics for a particular call, to the SSF/CCF. Upon receiving the IF, the SSF/CCF applies the charging procedure as requested and, in the end, sends the *Apply Charging Report* IF[178] to the SCF.

---

[175]This is the term used in the Recommendation. It refers to local exchange switches from which the calls are originated. The charging information is sent to them via pulses (so that the overall pulse count determines the charge for the call) or the Signaling System No. 7 Tariff messages.

[176]The Recommendation leaves the definition of such events to network operators. The examples (in Clause 6.4.2.34) include receipt of charging information (call tariff, tariff change, time of tariff change, number of pulses, etc.) from the called party side of the network.

[177]The one example of a charging procedure cited by the Recommendation in Clause 5.2.2.2.2 is the pulse generation procedure (see footnote 176). Other examples (given in Clause 6.4.2.6) include a tariff table, which the SSF/SCF should use to calculate the charge and the call tariff information itself.

[178]The information in the report is declared "network-specific." The Recommendation, however, mentions that it may contain, for example, the exact number of pulses issued for the call. There may be more than one report issued in response to one request.[179]This includes the information about the announcements to be played along with the charging information for these announcements, etc.

5.4.3.2.4   Compare SIB.   This SIB is realized entirely within the SCF.

5.4.3.2.5   Distribution SIB.   This SIB is realized entirely within the SCF.

5.4.3.2.6   Limit SIB.   This SIB may be realized only within the SCF (which would deny the IN treatment of some calls); another approach considered by the Recommendation is to realize the Limit SIB by using the information exchange over the SSF/CCF-to-SCF relationship *r3*. In the latter case, the SCF issues the *Activate Service Filtering* IF to the SSF/CCF, which specifies the time to start filtering, the duration of filtering (or, alternatively, the time it is to stop), the treatment[179] to be applied to the calls that have been filtered, the criteria for selecting the calls to be filtered,[180] and the algorithm (i.e., interval- or number-based call filtering) to be used. Upon reception of the SCF request, the SSF/CCF starts filtering. It issues the *Service Filtering Response* IF to the SCF every time the call is passed through (to receive an IN treatment). It also issues the same IF when the filtering is over.

5.4.3.2.7   Log Call Information SIB.   The SCF, SSF/CCF, and SDF are involved in the realization of this SIB. The relevant data are exchanged over the SSF/CCF-to-SCF relationship *r3,* and the SCF-to-SDF relationship *r4*.

The sequence of the execution of this SIB is as follows:

1. For a given call, the SCF issues the *Call Information Request* IF to the SSF/CCF. (The particular information requested may be any combination of the following items: the elapsed time since the call attempt had been made, time the call ended, the elapsed time since the connection[181] had been established, the calling and called numbers, and the bearer capability.) This request may be canceled by the SCF, in which case it needs to issue the *Cancel Call Information Request* IF to the SSF/CCF.

2. When the call is ended, the SSF/CCF responds—unless preempted by the cancellation—with the *Call Information Report* IF to the SCF.

3. The SCF logs the information into a file by issuing one or more of the following IFs to the SDF: *Add Entry, Modify Entry,* and *Remove Entry*.[182]

---

[180]These may be either specific destination or originating numbers or references to specific services.

[181]In CS-1, this SIB deals with two-party calls.

[182]The reader will not find all three IFs in Clause 5.2.6 (which provides the Stage 2 description of the *Log Call Information* SIB) of Recommendation Q.1214. Only *Modify Entry* is mentioned there; however, Clause 6.7 of the Recommendation specifically identifies all three as related to the realization of the SIB.[183]No nesting of SIBs is supported in CS-1 (CS-2 includes the work on the necessary constructs to support this capability).

4. The SDF responds to each SCF request with a response IF (*Add Entry Result, Modify Entry Result,* and *Remove Entry Result,* respectively).

5.4.3.2.8 Queue SIB. This is the most complex SIB, whose description is further complicated by the necessity to specify all the capabilities of the User Interaction SIB.[183]

In the previous chapter we described the messages exchanged between the switches and service control. It would be helpful if the reader consulted that material again to refresh his or her memory.

The SIB is realized through the SCF-to-SSF/CCF relationship $r3$ and SCF-to-SRF[184] relationship $r6$.

We recall that the issue here is keeping track within the SCF of the availability of the remote resources (i.e., lines or trunks associated with several terminating offices). While processing a query to the service control requesting a routing instruction, the SCF may determine that there is a free resource available, at which point it marks it as busy and issues the *Request Report BCSM Event* IF to the SSF/CCF. The event here is the disconnection of the call. When the call is disconnected, the SSF/CCF sends the *Event Report BCSM* IF (or a DP-specific report) to the SCF, which notifies the latter of the resource's available status.

If all resources are busy, the SCF queues the call. At this point, it does the following:

1. Issues the *Request Report BCSM Event* IF to the SSF/CCF, which asks the SSF/CCF to report the *abandonment* of the call by the calling party.[185]

2. If required, initiates playing announcements to the calling party (by the SRF). (All the procedures and message exchanges related to this activity are reviewed in Sec. 5.4.3.2.13.)

3. Issues the *Hold Call in Network* IF to the SSF/CCF and starts the internal timer whose value is smaller than the that used within the SSF/CCF to time a call that has not been connected. Whenever the SCF timer expires, it issues the *Reset Timer* IF to the SSF/CCF (which restarts its respective timer and thus prevents the destruction of the call), and restarts the timer again.

---

[183]No nesting of SIBs is supported in CS-1 (CS-2 includes the work on the necessary constructs to support this capability).

[184]The SRF is not *essential* for the description of this capability. (That is not to say, however, that it is not essential to the implementation of the capability!)

[185]If the call is abandoned, the SSF/CCF sends the respective *Event Report BCSM* IF to the SCF, and the SCF removes the call from the queue.

When the SCF receives the *Event Report BCSM* IF (or a DP-specific report) from the SSF/CCF and thus learns of the availability of the resource, it does the following:

- Terminates the announcements to the calling party at the head of the queue
- Advances the queue
- Issues the connection instruction to the SSF/CCF along with the *Request Report BCSM Event* IF

5.4.3.2.9 Screen SIB. This SIB is realized through the SCF-to-SDF relationship *r4*. The SCF sends the *Search* IF to the SDF. The IF, among other data, carries the name of the data object to be searched, aliases, and filtering information. The SDF responds with the *Search Result* IF, which echoes the search data augmented by the search result (*Match* or *No Match*).

5.4.3.2.10 Service Data Management SIB. This SIB is realized through the SCF-to-SDF relationship *r4*. The SCF may send one of the following IFs to the SDF: *Search, Modify Entry, Add Entry,* and *Remove Entry*. The SDF is to respond (after the operation has been completed) with, respectively, *Search Result, Modify Entry Result, Add Entry Result,* and *Remove Entry Result*.

5.4.3.2.11 Status Notification SIB. This SIB is realized over the SCF-to-SSF/CCF relationship *r3*.[186] The SCF issues the *Request Status Report* IF,[187] and it may also issue the *Cancel Status Report* IF. Depending on the request, the SSF/CCF responds with one or more *Status Report* IFs.

As we recall from the previous chapter, the four status notification capabilities defined in Recommendation Q.1213 are as follows:

1. *Polling* (i.e., checking the current status of the specified resource). This capability is achieved by specifying the value of the *Monitor Type* IE of the *Request Status Report* IF to *Poll Resource Status*.

2. *Waiting* until a specified resource changes its current status (either from *busy* to *idle* or vice versa). This capability is achieved by spec-

---

[186]Initially, i.e., in the 1993 version of Recommendation Q.1214, the SCF-to-SDF relationship *r4* was also involved. As the result of the CS-1 refinement work, the relationship *r4* was eliminated.

[187]One question, often asked at this point, is why the *Request Report BCSM Event* IF cannot be used here. The answer is that the resources monitored in this SIB are *physical* [a non-ISDN line, a directory number associated with an ISDN interface, a *Multi-Line Hunt Group* (MLHG), and a *Trunk Group* (TG) are the examples given in the Recommendation], while the BCSM resources are *logical*. In this respect, the capabilities of the present SIB belong to a lower level.

ifying the value of the Monitor Type IE of the Request Status Report IF to Monitor for Change.

3. Initiating *continuous monitoring*. This capability is achieved by specifying the value of the *Monitor Type* IE of the *Request Status Report* IF to *Continuous Monitor.*

4. Canceling continuous monitoring. This capability is achieved by issuing the *Cancel Status Report* IF.

5.4.3.2.12 Translate SIB. This SIB is realized through the SCF-to-SDF relationship *r4*. The SCF may send the *Search* IF to the SDF, which responds with the *Search Result* IF.

5.4.3.2.13 User Interaction SIB. The User Interaction capabilities are achieved in the DFP by establishing a connection between an end-user (call party) and the SRF.[188] The SRF receives the instructions from the SCF and plays the announcements or collects the user input (or both). If the user input is collected, it is subsequently sent to the SCF. It is clear that the User Interaction SIB should be realized over the SCF-to-SRF relationship *r6*.

There are two complications with that. The first complication is that, in CS-1, it is assumed that the relationship *r6* is actually supported via *relaying* the SCF messages over the relationships *r3* (SCF-to-SSF/CCF) and *r5* (SSF/CCF-to-SRF).[189] The second complication is that a network operator may choose to maintain different capabilities (i.e., specific announcements or voice recognition apparatus) in different physical incarnations of the SRF, attached to (or implemented within) different switches,[190] and therefore the network should be able to *handoff* the call to an appropriate switch, either temporarily or permanently.

We start with the first case, where the SRF with the needed capabilities is either

---

[188]This connection is currently assumed to be the bearer channel. Recommendation Q.1214 gives the only example of such a channel as a 64 kbit/s circuit, and adds that the use of an ISDN D-channel is for further study.

[189]This agreement took place as a compromise needed to reconcile three distinct types of implementations. In the first type, the SRF was placed inside the switches; the vendors of such products naturally preferred that the messages to the SRF be sent directly to the switches (i.e., over the *r3* relationship). In the second type, the SRF was placed within a stand-alone processor attached to a switch through a combination of a bearer and signaling connection. Finally [cf. the description of the Service Node (SN) PE in the next chapter], the SRF combined with the SCF and SSF/CCF in a single node, connected to a switch via an ISDN interface, makes the third type of implementation. As the reader can see, the common denominator of all three types of implementations is the relaying of the SCF-to-SRF communications through the switch.

[190]This was the experience with the AT&T network, for which the capabilities referred to later in this sentence were invented and first implemented.

1. Located within a switch that has an established relationship *r3* with the SCF (on behalf of a particular call)

2. Located in a separate entity that has a network connection *r5* with the switch that has an established relationship with the SCF

A typical scenario for this case is as follows. The switch requests an instruction (e.g., a Freephone number translation) via the *Initial DP* (or a DP-specific) IF to the SCF, at which point the relationship was established. The service logic then finds out that additional information has to be collected from the calling party. By examining the *SRF Available* and *SRF Capabilities* IEs that were carried by the IF, the SCF determines that it can use the SRF associated with the switch that issued the request for instruction.

To start the user interaction procedure, the SCF issues the Connect to Resource IF to the SSF/CCF over the *r3* relationship. From then on, until the end of the procedure, the SCF deals with the relationship *r6*, which happens to be mapped into either relationship *r3* or the {*r3, r5*} pair of relationships.[191] The user interaction procedure may be ended by the SRF (if the SCF specifically indicates that this may be the case in either of the two SCF commands discussed in the next paragraph), or the SCF ends it by issuing the *Disconnect Forward Connection* IF to the SSF/CCF.[192]

If all that is needed by the service is playing the announcements,[193] the SCF issues the *Play Announcement* IF.[194] Playing of the announcements can be canceled at any time by the SCF, which then issues the *Cancel Announcement* IF. If requested by the SCF, the SRF issues the *Specialized Resource Report* IF to the SCF when it finishes playing the announcements requested.

---

[191]Again, this means that the messages between the SCF and SRF are relayed by the SSF/CCF. Specifically in the case of the *Connect to Resource IF*, Recommendation Q.1214 allows, as an option, the subsequent exchange of the ISDN SETUP request and confirmation messages between the SSF/CCF and SRF to establish the connection.

[192]Recommendation Q.1214 allows, as an option, the subsequent sending of the ISDN RELEASE request by the SSF/CCF to the SRF to release the connection.

[193]Recall that Call Queuing is one important example of the feature that requires just playing announcements.

[194]Although, in general, we do not discuss IEs in this book, this case warrants the introduction of several IEs, if only to give the reader a feeling for the types of the announcement capabilities supported in CS-1. The *Information to Send* IE, for example, may, as an option, carry the text that is to be converted to speech by the SRF. (Thus, instead of keeping the announcements in the SRF entities themselves, network operators can keep them at central databases stored as text, which occupies comparatively little memory.) The same parameter carries the number of repetitions for the announcement to be played. The *Disconnection from IP Forbidden* IE indicates whether the SRF may disconnect when it is done playing the present announcement or wait for another instruction. The *Request Announcement Completed Indication* IE indicates whether or not the SRF is to reply with the report after it finished playing the announcements.

If the service logic needs to collect information from the user, the SCF issues the *Prompt and Collect User Information* IF to the SRF, to which the SRF replies with the *Collected Information* IF.

This completes the description of the user interaction procedure for the case in which the SRF is either located within or attached to the switch that has an established relationship with the SCF at the moment the user interaction is required.

The second case is when the switch that has an established relationship with the SCF either has no SRF[195] or the capabilities of the SRF it has are insufficient as far the requirements of the service are concerned. Given that there is another SRF with the capabilities needed and that the SCF knows the address of the switch associated with it, the SCF may start one of the two procedures that connect the call party with the SRF that has the desired capabilities. The names of these procedures are *Service Assist* and *Service Hand-off,* with the latter being a subset of the former, for which reason both procedures are often referred to by a common name *Service Assist / Hand-off.* We describe the procedure with the aid of the Fig. 5.12.

The SSF/CCF that is associated with the SRF with desired capabilities is called the *assisting* SSF/CCF. The SSF/CCF that has the initial

---

[195]That is, it has no internal SRF and no attached SRF.

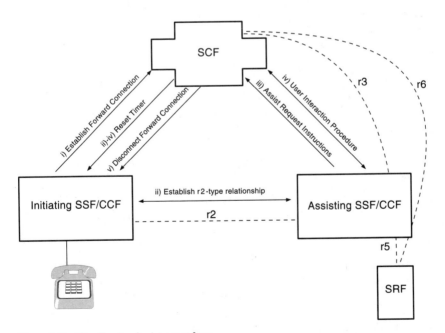

Figure 5.12   The *Service Assist* procedure.

relationship with the SCF is called the *initiating* SSF/CCF. (For simplicity, Fig. 5.12 depicts the call party attached to the initiating SSF/CCF. This may not be the case though, especially if the initiating switch is the toll switch.) The procedure starts with (1) the SCF issuing the *Establish Temporary Connection* IF over the relationship *r3* to the initiating SSF/CCF. This IF carries the network address of the assisting SSF/CCF. Then (2) the initiating SSF/CCF establishes the *r2*-type relationship with the assisting SSF/CCF and signals to it to request further instructions from the SCF.[196] The assisting SSF/CCF, in turn, (3) issues the *Assist Request Instructions* IF to the SCF, at which point the relationship *r3* is established between the SCF and the assisting SSF/CCF. Then (4) the SCF and the assisting SSF/CCF perform the user interaction procedure as described above. In parallel with steps (2) through (4), the SCF issues the *Reset Timer* IF to the initiating SSF/CCF in order to prevent the timeout of the call that, as far as the initiating SSF/CCF is concerned, has not yet been terminated to another call party. The Service Assist procedure is terminated by the SCF, which (5) issues the Disconnect Forward Connection IF to the Initiating SSF/CCF, followed by the instruction on how to complete the call.

The Service Hand-off procedure differs from the Assist one in that the former omits step (5).[197]

Finally, Recommendation Q.1214 considers a special case in which the SRF itself (i.e., without an SSF/CCF acting as an intermediary) is capable of establishing the *r2*-type relationship with the initiating SSF/CCF. In this case, after this relationship is established, the SRF establishes the *r6* relationship with the SCF by issuing to it the Assist Request Instruction from SRF IF.

5.4.3.2.14 Verify SIB. This SIB is realized entirely within the SCF.

5.4.3.3 Other service-independent capabilities. As we have pointed out several times, the SIBs are the instructions available to a service programmer. Additional instructions are needed—and usually made available—for programming operating systems. The IN, as it grows, assumes the role of an operating system that controls the network and provides the environment for executing service applications.

Certain capabilities, such as network management, although auxiliary to service introduction *per se,* are essential to proper operation of

---

[196]Normally, this is done via ISUP protocol, and this is one example where the ISUP/INAP interworking is necessary to achieve harmony in service introduction.

[197]The standard stops short of prescribing what exactly may be done next in the case of Service Hand-off. Presumably, the assisting SSF/CCF can inherit the state of the initiating SSF/CCF at the time the latter sends the initial DP (or a DP-specific) IF so that the SCF may issue the instruction to the assisting SSF/CCF. In the AT&T network, the Service Hand-off capability has been used only for playing terminating announcements.

the network. These capabilities have been studied and discussed in standards for several years; two of them, called *Activity Test* and *Call Gapping,* have been standardized in CS-1.

5.4.3.3.1 Activity Test capability. This capability allows the SCF to test the existence of the network connection to the SSF/CCF. Recommendation Q.1214 uses the word "relationship" here, although one may argue that this word should be reserved for the abstract connection created on behalf of a particular call. In other words, there may be an existing connection between the SCF/SSF over which several relationships exist. One particular relationship may cease to exist (because a process, say, in the SCF crashed), but the network connection still exists. We will follow the terminology of the Recommendation; however, when we use the word "relationship," we should be aware that the meaning of it changes with the context.[198]

The *Activity Test* is performed when the SCF has to ascertain whether it is possible to exchange messages with the SSF/CCF. This may be needed when the associated computer equipment is rebooted or when there has been no traffic between the two FEs for some time.

To perform this capability, the SCF issues the *Activity Test* IF[199] over the SCF-to-SSF/CCF relationship *r3* and waits for the *Activity Test Response* IF from the SSF/CCF for a period of time specified by the network operator. If the response does not arrive within the specified period of time, the test is considered failed.

5.4.3.3.2 Call Gapping capability. This capability allows control over the overall overload of the SCF. The service filtering capability of the *Limit* SIB is very similar to this one. There are two major differences:

1. Call Gapping passes one call during a specified time period, the *gap.*

2. No response is to be sent from the SSF/CCF (contrary to the case of service filtering).

The SCF issues the *Call Gap* IF to the SSF/CCF, which specifies the duration of gapping, the treatment[200] to be applied to the calls that have been filtered, the criteria for rejecting the calls,[201] and the gap interval.

---

[198]The user may have already observed that the "relationship" of the Limit SIB is also non-call-related.

[199]This is one IF that carries no IEs.

[200]This includes the information about the announcements to be played.

[201]As in the case of service filtering, those may be either specific destination or originating numbers or references to specific services or even DPs.

5.5   Draft Recommendation Q.1224,
Intelligent Network—Distributed Functional
Plane for CS-2 (Highlights)

5.5.1   Overview

The functional capabilities at the DFP for CS-2 currently include the following:

1. *End-user access to call and service processing,* via the analog line interfaces, ISDN BRI and PRI, and traditional trunk and SS No. 7 interfaces.[202]

2. *Service invocation and control.* The single-endedness and single-point-of-control principles of CS-1 are still the "law" of CS-2. And, as it was in CS-1, the interface between the SSF and CCF is not open in CS-2. CS-2, however, does support calls that involve more than two users.

3. *End-user (i.e., call-party) interaction with the network.* End-user interaction with the network to send and receive information is provided by service switching and call control resources, augmented by specialized resources. This faculty—already existent in CS-1—is further augmented in CS-2 by (1) standardization of the capabilities (such as *User-Interaction Scripts*)[203] that enhance the network performance when using the SRF and (2) standardization of the SRF capabilities for call bridging.[204]

4. *Service management.* The service management capabilities regarding the provisioning and maintenance of the service logic and data are described in CS-2 in much more detail than in CS-1; however, no interfaces have been standardized in CS-2. Nevertheless, the capability of service subscribers to change the subscriber-specific service management information is supported in CS-2.

5. *Call Party Handling* (CPH). CPH enables IN to influence multiparty calls, and it satisfies the needs of benchmark services such as Conference Calling, Call Hold, Call Transfer, and Call Waiting.

6. *Internetworking.* For the internetworking purposes, IN CS-2 identifies the SCF-to-SCF, SMF-to-SMF, and SDF-to-SDF relationships in

---

[202]As the reader can see, these particular capabilities are the same as in CS-1.

[203]The *User Interaction Scripts* contain the part of the service logic that deals with the specific interaction with the call party. In contrast to CS-1, where the SCF is to send *one* instruction at a time to the SRF to play an announcement and collect digits in response to the announcement, in CS-2 the user-interaction script (i.e., a program governing playing multiple announcements, collecting responses to them, and then carrying on based on such responses) is executed by the SRF.

[204]See the description of the *hybrid* approach to CPH later in this section.

addition to the IN CS-1 SCF-to-SDF relationship. In accordance with the single-point-of-control principle of CS-2, however, distributed service control is not supported.[205]

7. *Security.* Secured data acquisition is a CS-2 requirement for both the SDF-to-SDF and the SCF-to-SDF relationships.

8. *Wireless access.* IN CS-2 currently supports both the wireline (fixed) and wireless access by users, but only at the DFP level. In order to meet the needs of wireless activities, out-of-channel user interaction (both call-related and call-unrelated) is supported in CS-2.

The rest of this section highlights the CS-2 FEs and relationships (with the emphasis on the wireless FEs), the CS-2 call model, and the CPH, in that order.[206]

5.5.2   Functional Entities and Relationships

The definitions of the CS-2 FEs are divided between the main text of the Recommendation and its Annex B. The division takes place among those FEs—and relevant relationships—that are not specifically related to the wireless aspects of telecommunications and those that are. The former are placed in the main text of the Recommendation, while the wireless-specific aspects are assigned to Annex B. Figure 5.13 depicts *all* present CS-2 FEs, for which reason it combines the FEs and relationships defined in the main text of the Recommendation with those defined in Annex B.

The following two sections are, respectively, dedicated to both types of the FEs.

5.5.2.1   New CS-2 FEs defined in the main text of the Recommendation.

These three new FEs are the *Intelligent Access Function* (IAF), *Call-Unrelated Service Function* (CUSF), and *Service Control User Agent Function* (SCUAF). In addition, the SRF capabilities have been enhanced in CS-2.

The *Intelligent Access Function* (IAF) provides access to IN-structured networks from non-IN-structured networks. (The IAF is actually located in a non-IN-structured network.) As far as the IN-

---

[205]Although many aspects of internetworking are unstable at the moment, much work has been carried out in that area. It is possible that by the time CS-2 is finished, it will support internetwork management interactions.

[206]The authors have to repeat once more that the work on the CS-2 material had *not* been finished by the time this book was submitted to the publisher. For this reason, only the material that, on the one hand, constitutes in the authors' opinion significant departure from CS-1, and, on the other hand, has been declared stable by the editors is discussed in this book.

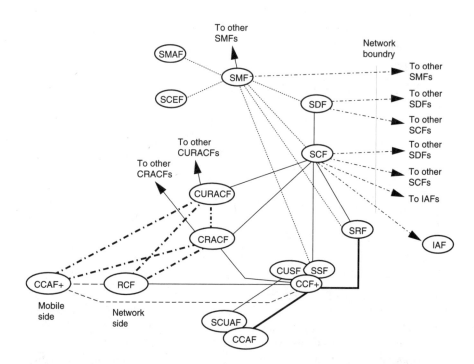

LEGEND

CCAF	Call Control Agent Function
CCAF+	Call Control Agent Function (Plus)
CCF+	Call Control Function (Plus)
CRACF	Call/connection related Radio Access Control Function
CUSF	Call-Unrelated Service Function
IAF	Intelligent Access Function
CURACF	Call-Unrelated Radio Access Control Function
RCF	Radio Control Function
SCEF	Service Creation Environment Function
SCF	Service Control Function
SCUAF	Service Control User Access Function
SDF	Service Data Function
SMAF	Service Management Access Function
SMF	Service Management Function
SRF	Specialized Resource Function
SSF	Service Switching Function

··················	Management Relationship
———————	IN Service Control
▬▬▬▬▬▬	Bearer Connection Control
– – – – –	Non-IN Call Control
▪—·—·—▪	Non-Call, Non-Bearer Associated IN Service Control and Radio Bearer Associated Control
–·–·–·–·–	Internetworking Relationship
– – – – – –	Radio Bearer Connection Control

Figure 5.13 IN CS-2 DFP Architecture (including mobility aspects).

structured network is concerned, the only FE there that has the relationship to the IAF is the SCF. The role of the IAF is pretty much that of a protocol converter.

The *Call-Unrelated Service Function* (CUSF), which is coupled with the SSF and CCF, supports the call-unrelated interactions between users and service processing.

The *Service Control User Agent Function* (SCUAF) provides the user access to the CUSF (much like the CCAF provides the user access to the SSF/CCF).

5.5.2.2 New CS-2 FEs defined in Annex B (wireless FEs).

Annex B of the Recommendation adds the wireless aspect of IN by building "upon the IN CS-2 functional architecture to provide guidance and direction as to how IN CS-2 can be used as the basis for wireless systems." As the Recommendation states, the Annex "is intended to impart to the reader insight as to the manner in which the IN principles can be applied to wireless access."

The wireless-access-related FEs are the *Call-Related Radio Access Control Function* (CRACF), *Call-Unrelated Radio Access Functions* (CURACF), *Radio Control Function* (RCF), and *Call Control Agent Function Plus* (CCAF+).

The CRACF supports service features and signaling that require handling of radio links. The CRACF supports the wireless-specific interactions within an IN-structured network. These interactions include *terminal paging,*[207] *radio-bearer setup,* and *handover.*[208] In addition, the CRACF allocates specific network radio system elements and other network resources for use during calls, and it may support the acquisition of current terminal location information within the local radio system environment, to support delivery of calls to the terminal. The CRACF is central to IN in that it is the CRACF that recognizes the need for IN processing and passes the call-related events and information to the SCF or SSF/CCF.

Annex B defines the *Terminal Access State Model* (TASM), which models the CRACF activities required to establish and maintain a terminal access from a mobile terminal to the network. As the Recommendation puts it, "the TASM is primarily an explanatory tool for providing a representation of CRACF activities that can be analyzed to

---

[207]In wireless services, the mobile terminal has to be located (before the call is delivered to it) in order to determine its routing address. Terminal Paging is the procedure performed to find the cell location where a mobile terminal resides. Normally, the SCF requests a CRACF to page the mobile terminal (with the *Page* IF) and send back the result (in the *Page Response* IF) when the terminal is found. If the response from the mobile terminal is not received within a certain period of time, the CRACF informs the SCF about that (with the *Page Response Not Received* IF). The SCF may start the paging procedure with several CRACFs and after receiving the first response cancel it (via the *Cancel Page* IF) at the rest of the CRACFs.

[208]Handover is the terminal mobility procedure, which reroutes a call to a new base station. In most present systems, handover is used to satisfy the speech and data quality requirements. Theoretically, the procedure could be also used for reasons unrelated to the link quality. Thus, handover could be initiated by the call party (for subscription-related reasons) or by a network (for network-management purposes).

determine which aspects of the TASM will be visible to IN." As such, the authors found the TASM both useful and self-explanatory. Figure 5.14 reproduces the Annex figure (which is now labeled "yyy").

The CURACF, as its name implies, is the non-call-related counterpart of the CRACF. The CURACF supports interactions within an IN network prior to a call reference being established or for which there is no type of relationship to a particular call. (Among the most important

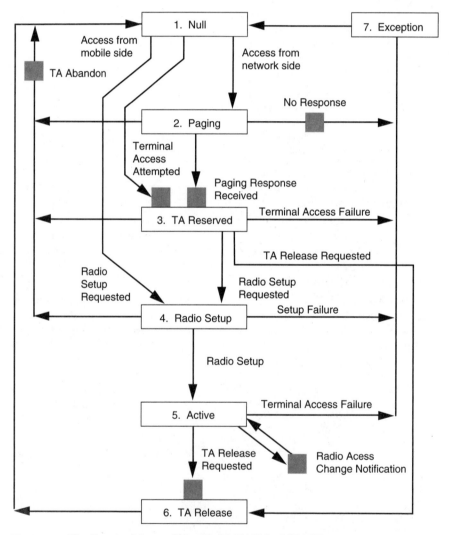

Figure 5.14   The Terminal Access State Model (TASM) of CRACF.

responsibilities of the CURACF is handling of all mobile-specific aspects of charging. Overall, the CURACF is responsible for all aspects of user registration and user authentication.) The CURACF maintains the radio-system side of communications with the SCF for all call-unrelated service features; it also administers the relationship between the CCAF+ and the network for the non-call-related interactions between users and service processing. Similarly to CRACF, CURACF is responsible for recognition of IN triggers and passing them to the SCF. Note, however, that unlike the CRACF the CURACF has no relationship with the SSF/CCF. The Annex text indicates that "it is possible that a handover situation may occur during the call-unrelated interactions in which only the CURACF is participating...[in which case it] must coordinate the handover with the CRACF...."

The non-call-associated interactions performed by the CURACF are represented in the *Basic Non-Call-Associated State Model* (BNCSM). Figure 5.15 (after the figure currently labeled "xxx" of Annex B) depicts the BNCSM.

The RCF establishes, maintains, modifies, and releases both radio and fixed-line bearer connections to the network. The Recommendation motivates the introduction of the RCF as a catchall for the functional capabilities that are not supported by CRACF and CURACF. To this end, the RCF's responsibilities include ensuring the security and privacy of users. The RCF receives indications from the terminals and, depending on their nature, relays them to either CRACF or CURACF. It also reports to the latter the state of the radio link connections between the terminals and network.

The CCAF+ is based on the CCAF, but it has an additional capability for both call-related and call-unrelated access to wireless terminals. The CCAF+ accesses the capabilities of the *Call Control Function Plus*[209] d(CCF+), RCF, and CRACF, using service requests (e.g., *setup, transfer, hold,* etc.) for the establishment, manipulation, and release of a call or an instance of a call-related service. Similarly, the CCAF+ accesses the capabilities of the RCF and CURACF for call-unrelated service interactions.

In addition to the IFs for the paging procedure (see footnote 207), the Annex defines the following IFs related to wireless:

- IFs between the SCF and CRACF:

    Terminal Access Attempted (issued by the CRACF to request authorization of the mobile terminal's access to the network)

---

[209]The word "plus" indicates that this FE may be required to be capable of more things than a normal CCF is (because it has, for example, to communicate with the CCAF+).

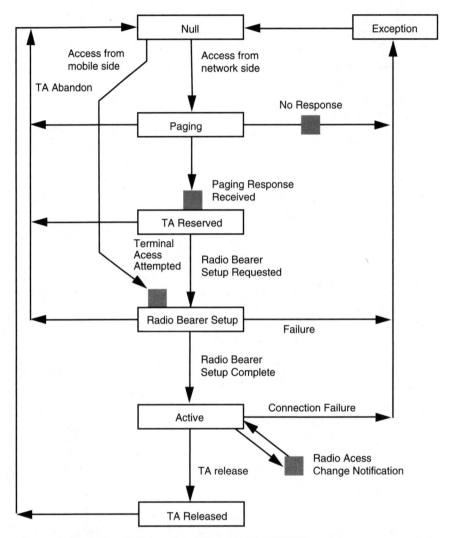

Figure 5.15   Basic Non-Call-Associated State Model (BNCSM).

Terminal Access Authorized (issued by the SCF in response to the request)

Terminal Access Release (issued by the SCF in order to instruct the CRACF to release all mobile-terminal-related resources related to the call)

Terminal Access Release Requested (issued to the SCF as a report on the call release attempt)

Wireless Call-Related Component Received (issued by the CRACF as a report on the reception of the specified component from the CCAF+)

Wireless Call-Related Send Component (issued by the SCF as a request to send a specified component to the CCAF+)

- IFs between the SCF and CURACF:

Wireless Call-Unrelated Activation Received (issued by the CURACF as a report on the association request)

Wireless Call-Unrelated Component Received (issued by the CURACF)

Wireless Call-Unrelated Send Component (issued by the SCF)

5.5.3   The CS-2 Call Model

As the Recommendation explains, the BCSM for IN CS-2 (described in Clause 4.2.2 of the present text of the Recommendation) "is based on the overall BCSM in Annex A/Q.1204 and Q.1214, refined as applicable to IN CS-2." As in CS-1, it is described with the two state machines, the Originating BCSM (depicted in Fig. 5.16) and the Terminating BCSM (depicted in Fig. 5.17).

In what follows, we compare the BCSM of Recommendation Q.1224 to those defined in Recommendations Q.1204 and Q.1214.

The reader will find both the originating and terminating models appear to be similar to those of Recommendation Q.1204 (cf. Figs. 5.5 and 5.6, respectively) in that they have the same number of PICs.

There are notable differences, however. Starting with the cosmetic ones, note that PICs and DPs are no longer assigned numbers.[210] More substantial modifications to the call model are manifested in the new functions of the PICs and DPs and their applications to services.

As far as the PICs go, the O_Disconnect and T_Disconnect PICs have been replaced in the originating and terminating models by, respectively, the O_Suspended and T_Suspended PICs. Note that this is not simply a name change, nor is it an arbitrary modification.

In a nutshell, the justification for the changes is dictated by a special way in which many services are set up [starting with the *Plain Old*

---

[210]The reason for this is that it became difficult to maintain these numbers so that the numbering is backward-compatible. To appreciate this, consider what has to happen when a DP or PIC is to be erased in the new version of the standard. In this case the standard is to end up with either a "hole" in the numbers or having all DPs whose numbers are above the erased one be renumbered. The former solution is somewhat unaesthetic; the latter one could be dangerously confusing.

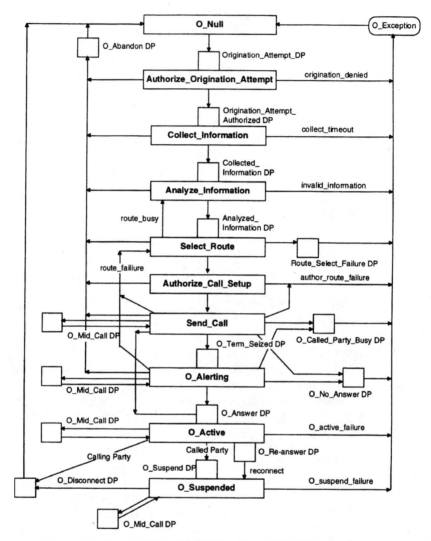

Figure 5.16   Originating BCSM for CS-2. (*After Fig. 4-3/Q.1224.*)

Telephone Service (POTS) in the United States]: unless the called party explicitly wishes to disconnect the call, it should not be immediately disconnected. The case is asymmetric (i.e., the originating and terminating parties are dealt with differently) for the following reason.[211] If the called party picks up a telephone set in an inconvenient

---

[211]This reason has been recognized and supported by the U.S. implementation since the old Bell System times.

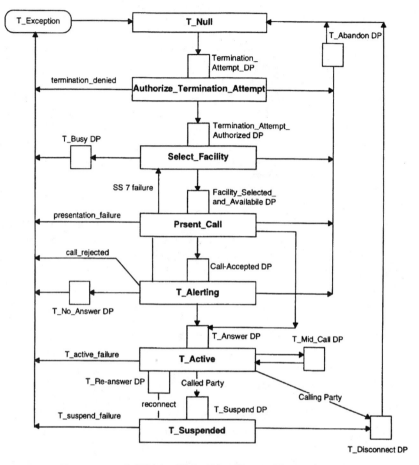

Figure 5.17 Terminating BCSM for CS-2. (*After Fig. 4-4/Q.1224.*)

place (say a kitchen) but would like to continue the conversation using another telephone in a more suitable place, the party is able to hang up momentarily (typically, for 10 seconds) while moving to another location in the house. This capability is called *reanswering.*

Naturally, in order to reanswer the call, it cannot be disconnected, even though the called party did hang up. The state of the call is captured in the model as *suspended,* and it gets special treatment. While the call is suspended, both the calling and called parties are placed in the *O_Suspended* and *T_Suspended* PICs. The Recommendation states that the *O_Suspended* PIC is entered when "a suspend indication is received from the *T_BCSM* when the terminating party has disconnected...." (The *suspend* indication is sent by *T-BCSM* when the terminating party hangs up.) On the terminating side, the physical resources associated with the call remain connected and the appropriate timer is started. If the terminating side reanswers before the timer has run out, *T_BCSM* sends the *reanswer* indication to the *O_BCSM*, and the call becomes active again (the originating and terminating parties are reconnected). Otherwise, *T_BCSM* sends the *disconnect* notification. The Recommendation is careful to indicate a difference in handling an SS7 trunk connection or ISDN interfaces on the one hand, and the analogue ones on the other. In the latter case, hanging up is ambiguous in that it may mean either an intent to reanswer or to end the call. In order to resolve this ambiguity, a timer is started when the called party goes on-hook. In the case of a trunk connection, there is actually no ambiguity: if the call is to be disconnected, the Q.931 *Disconnect* message (in case of the ISDN connection) or the ISUP *Release* message (in case of a trunk connections) is received on the terminating side, which makes the *T_BCSM* immediately recognize that the call is to be disconnected; if the call is to be suspended, however, that would be indicated by different messages. But even if the terminating party disconnects, the originating party may wish to make another call without repeating any authorization procedure (i.e., reentering the calling card number), for which reason the *O_BCSM* supports the *O-Mid-Call* DP and the respective transition from the *O_Suspended* PIC.

As far as the CS-2 transitions go—even though they almost mimic those of Recommendation Q.1204—fewer of them (compared to the example BCSM) are associated with DPs. In other words, not all DPs defined in Recommendation Q.1204 are supported in CS-2. Of the transitions leading to exception handling, only *O_Route_Select_Failure, O_Called_Party_Busy, O_No_Answer, O_Mid-Call, T_[Called_Party]_Busy, T_No_Answer, T_Disconnect,* and *T_Mid_Call* have become the DPs in CS-2. The *Route_Selected* and *Origination_Authorized* transitions in *O_BCSM* are not associated with the CS-2 DPs either. On the other hand, CS-2 is enriching the example model by adding new DPs: *O_Suspend, O_Reanswer, T_Suspend,* and *T_Reanswer.*

Even though the IN capabilities modeled in BCSM have not reached the level of the general model of Recommendation Q.1204, they are substantially advanced compared to those of CS-1, as the reader can observe by returning to Figs. 5.9 and 5.10.

To conclude this section, we note that Recommendation Q.1224 defines the full set of CS-2 transitions—both the IN-supported ones and those supported only within the basic call. The transitions beyond the basic call cannot be inferred from the BCSM model, for which reason we provide them in the following two lists, one for *O_BCSM* and one for *T_BCSM*, which relate each DP to the set of PICs reachable from it by the beyond-basic call. To emphasize the nontrivial aspects of the IN processing, the lists refer *only* to the transitions that are not demonstrated in Figs. 5.16 and 5.17.

The *O_BCSM* DPs may result in the IN transitions to the following PICs:

- *Origination_Attempt: Collect_Information, Analyze_Information,* and *Select_Route*
- *Origination_Attempt_Authorized: Analyze_Information* and *Select_Route*
- *Collected_Information: Collect_Information* and *Select_Route*
- *Analyzed_Information: Collect_Information* and *Analyze_ Information*
- *Route_Select_Failure: Collect_Information, Analyze_Information,* and *Select_Route*
- *O_Called_Party_Busy: Collect_Information, Analyze_Information,* and *Select_Route*
- *O_No_Answer: Collect_Information, Analyze_Information,* and *Select_Route*
- *O_Mid_Call: Analyze_Information, Select_Route, O_Suspended, Collect_Information, Analyze_Information,* and *Select_Route*
- *O_Disconnect: Analyze_Information* and *Select_Route*

The *T_BCSM* DPs may result in the IN transitions to the following PICs:

- *Termination_Attempt: Select_Facility*
- *T_Busy: Present_Call*
- *T_No_Answer: Select_Facility*

5.5.4 Call Party Handling

Call Party Handling (CPH) enables IN to deal with multiparty calls. CS-2 accepts the *Connection View* model (see Sec. 5.2.5.3) and defines the operations on the individual legs of a call. The latter are brought from the appendix of Recommendation Q.1214. In addition, CS-2 presents a new concept of *Connection View States* (which, until recently, were referred to as *Call Configurations*). The concept has been developed into a mechanism that provides the same level of IN support of multiparty calls as the individual leg-related operations do, but at much less cost to switch developers, and with fewer messages to cross networks at the run time.

Not every switch can support the IN CPH. The Recommendation defines four *Core Capabilities* required at the SSF/CCF to support CPH, which are as follows:

1. *Core Capability 1* allows a call party to initiate midcall interactions and provide additional input, which is then relayed to the SCF. Thus, the support of the Mid-call DP is a necessary condition for providing this capability. As far as the interfaces are concerned, the Recommendation mentions the following possibilities: "Possible support of this capability can be through the use of display phones with feature selection, use of SRFs to play announcements and to collect digits, or by offering dial tone to the user for the input of service codes. Although various possibilities exist to support this capability, that which is standardized should lend support to both analog as well as ISDN BRI and PRI access," noting, however, that the use of analogue trunks is left for future study.

2. *Core Capability 2* allows each call party to be connected to a resource as well as to transfer a call to another party. This, in turn, requires that the SSF/CCF provide to the network the information necessary to correlate the independent transport paths and the intended destinations.

3. *Core Capability 3* is the ability to present the current half-call view to the SCF. The Recommendation clarifies this requirement as follows: "The SCF should be able to observe the present status of the call at a given SSF/CCF based on the originating or terminating half call model. This view should be sufficient for service logic within the SCF to be able to determine the status of current connections to the *controlling*[212] leg. This view should be able to provide information for each call party relating to the Basic Call State Model (BCSM), as well as the current conditions or events."

4. *Core Capability 4* is "the ability for the SSF/CCF to combine selected transferred paths within the SSP into a call. A mechanism must exist that will allow the SSF/CCF to regain control of the individual transport paths of the call parties, after a transfer or connection to resource has been made. Upon notification from IN service logic, the SSF/CCF is to combine selected individual paths within the SSP into a call on behalf of the controlling leg."

The individual-leg-related operations and the Connection View States approach are presented in the following two sections.

---

[212]Because of the single-endedness principle of CS-2, the service logic may control (or at least be aware of controlling) only one leg in a call. Thus, as far as the SCF is concerned there is only *one* controlling leg.

5.5.4.1 Individual-leg-related operations. The operations that deal with the individual legs (and segments) are represented by the following IFs: Initiate Call Attempt, Release Call, Connect, Disassociate Call Segments, Disconnect Leg, Merge Call Segments, Move Call Segments, Reconnect, and Split Leg.

The first three IFs are—at the level of detail addressed in this book— the same as they where in CS-1. The new IFs are as follows:

The *Disassociate Call Segments* IF is an instruction to the SSF/CCF to end the association between two call segments, thus creating two different calls. The SCF has an option to assign correlator IDs to the new calls so as to reference them in the future.

The *Disconnect Leg* IF is an instruction to the SSF/CCF to release one leg in the call (but retain the rest of the legs and the call they are comprising intact).

The *Merge Call Segments* IF is an instruction to the SSF/CCF to merge two associated call segments into one call segment. The execution of the instruction results in establishing a communication path connecting the (single) controlling leg of the associated segments with the rest of the legs.

The *Move Call Segments* IF is an instruction to the SSF/CCF first to offer the call to the *busy* called party and then (i.e., after the called party accepts it) bridge the two call segments into a single call segment. This instruction supports the capability to join a conference call in progress at the discretion of the conference chair.

The *Reconnect* IF is an instruction to the SSF/CCF to reestablish the communication path between the controlling leg (which had previously disconnected a call after putting its interlocutor on hold) and the held party. This operation is performed only when the party attached to the controlling leg does not have an ISDN interface.

The *Split Leg* IF has an effect that is the inverse of the Merge Call Segments IF. It is an instruction to the SSF/CCF to separate a particular leg from its call segment and place it into a new call segment.

5.5.4.2 Connection View States (formerly Call Configurations). The concept of the Connection View States (CVSs) (which were initially called Call Configurations[213]) was introduced to standards in Isidoro (1992).

---

[213]Unfortunately, the standards participants change the names of things much too often. Although there is merit in the term *Connection View States*—it appropriately reflects the fact that each call configuration is a view of the state of the call (combining all its connections) provided by the SSF/CCF to the SCF, the name is awkward. In what follows, we use the old term, *Call Configuration,* and the new one interchangeably.

It is based on the observation that for all practical purposes there are only so many ways in which the call parties can be connected to each other. Any call can be viewed as a progression of such constellations of the connections (or configurations), from the empty one (with no connections at all) to the one that includes one party (originating a call) to the one that includes two parties, etc., back to the empty configuration (when all parties hang up). The configurations can be catalogued and enumerated. Then, during the execution of a call, the SCF can send one instruction to the SSF/CCF to move to a desirable configuration instead of issuing multiple microinstructions for each leg involved. When fully specified and implemented, this approach will significantly improve the service performance (while reducing the load on the signaling network), and it will also reduce the complexity of the service logic and practically eliminate potential run-time errors that could seriously disable a switch.

In CS-2, this approach has not been carried to the point where the protocol messages carrying specific SCF instructions that force transitions from one CVS to another are defined—the protocol specifies only "micro" instructions; however, the CVSs currently defined *limit* the otherwise infinite set of configurations (some of which are potentially dangerous) to a well-known subset, which supports the configurations required by most (if not all) CS-2 services. Furthermore, CS-2 demonstrates in terms of the CVSs how the individual leg operations can be used to influence calls, thus providing a discipline for the use of the individual leg operations.

The Recommendation defines each CVS in terms of its relationship to BCSM, the entry events (i.e., the events that led to establishing the configuration reflected in the state), and the exit events (i.e., the events that result in a transition to another CVS).

Figure 5.18 demonstrates the conventions used in graphical representations of CVSs. The objects depicted are the legs (labeled with their IDs, in which prefix "c" indicates the controlling leg and prefix "p" indicates a passive leg), connection points, call segments, and their associations.

Figure 5.19 accompanies the discussion of the present catalogue, which contains the following 12 CVSs:

1. The *Null* CVS corresponds to an initial state where call processing is inactive. (Thus, only a connection point is represented, but no legs are attached to it.)

2. The *Originating Setup* CVS corresponds to an originating two-party call in the setup phase.

3. The *Stable 2-Party* CVS corresponds to a stable two-party call (either an originating or terminating half-call).

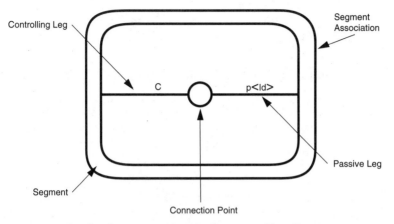

Figure 5.18 Graphical representation of a Connection View State.

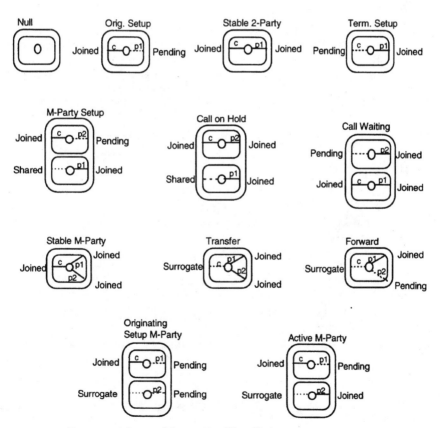

Figure 5.19 Present catalogue of Connection View States.

4. The *Terminating Setup* CVS corresponds to a terminating two-party call in a setup phase.

5. The *M-Party Setup* CVS[214] corresponds to the situation where the party attached to a controlling leg has put one party on hold while originating new calls.

6. The *Call on Hold* CVS corresponds to a call with two associated segments, where the party attached to the controlling leg is participating in a call (with the party attached to another active leg) after having put on hold the party attached to a passive leg.

7. The *Call Waiting* CVS also corresponds to a call with two associated call segments. The party attached to the controlling leg is participating in a call that is in the stable (or clearing) phase, while another call is waiting before it can terminate to the party attached to the controlling leg.

8. The *Stable M-Party* CVS corresponds to a stable (or clearing) multiparty call.

9. The *Transfer* CVS corresponds to a stable call that has been transferred by a party attached to the controlling leg. The controlling leg is labeled "surrogate" in reference to the billing relationship between the two passive legs after the call has been transferred.

10. The *Forward* CVS corresponds to a call forwarded by the party attached to the controlling leg.

11. The *Originating Setup M-Party* CVS corresponds to the situation in which the party attached to the controlling leg is setting up a call.

12. The *Active M-Party* CVS corresponds to the situation when the multiparty call has reached a stable active state (i.e., on detection of the Call Setup Authorized event in the BCSM of the last pending connection).

Figure 5.20 demonstrates the transitions among CVSs. (The present text of the Recommendation contains a complete list of the individual-leg-related operations necessary to achieve the transitions.)

To provide more complex CPH services, the hybrid approach is defined. This approach combines the (diminished) capabilities of an SSF/CCF regarding the CVSs with a bridging SRF in an external platform. While the hybrid approach requires fewer CVSs to be supported

---

[214]"M-Party" here and in the rest of this section stands for *Multi-Party*. (The respective figures demonstrate three parties involved in such calls.)

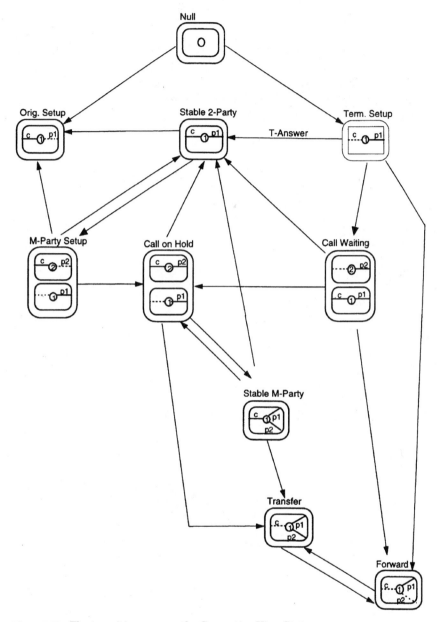

Figure 5.20 The transitions among the Connection View States.

by the SSF/CCF, more capabilities are needed in the SRF. For this reason, CS-2 adds such capabilities.

In CS-1, the SRF connections can only be made when call processing is suspended. In CS-2, however, the SRF connections can be made at any time, concurrently with call processing. In this case, the SRF can be used to recognize the signals from the user supplied via telephone dials as well as the voice.

Chapter

6

IN Physical Plane

6.1 Overview

There are two aspects to the Physical Plane material. One aspect is the definition of the Physical Entities (PEs) and the assignment of the Functional Entities (FEs) of the distributed functional plane to these PEs. This aspect is covered in Recommendations Q.1205 (general issues), Q.1215 (CS-1-specific issues), and Q.1225 (CS-2-specific issues). The second aspect is the specification of the Intelligent Network Application Protocol (INAP), which is defined—to achieve flexibility—for the interfaces among the FEs rather than PEs. This aspect is covered in Recommendations Q.1208, Q.1218, and Q.1228. In addition, Recommendations Q.1211 and Q.1221 respectively define the scope of the CS-1 and CS-2 protocol specifications.[215]

The protocol Recommendations are founded on the considerable amount of material related to the application layer standards; some of this material forms a prerequisite for understanding the INAP. The authors have attempted to provide a brief tutorial as part of the review of the protocol Recommendations on the essential aspects of such material.

This chapter is dedicated to Recommendations Q.1205, Q.1208, Q.1211, Q.1215, Q.1218, and the present status of the Draft Recommendation Q.1225, with one section for each Recommendation.

---

[215]At the time of writing this book, Recommendations Q.1221, Q.1225, and Q.1228 still exist only as drafts. Even though the authors discuss only the material they believe is stable, the reader is urged to remember that this material is only a candidate for standardization, not a standard, and that it may change considerably or even not be included in the final text of the Recommendations.

But before proceeding with these, we owe the reader an explanation for omitting two draft CS-2 Recommendations. The reason is that neither the material of Recommendation Q.1221 nor that of Recommendation Q.1228 is sufficiently stable at this moment to warrant a detailed discussion. Presently, Recommendation Q.1228 contains over 1000 pages (which makes it much larger than the final draft of the whole CS-1 in 1993!), of which quite a few are dedicated to the issues that are still questioned. Perhaps the newest and most interesting development in Recommendation Q.1228 is the support of *security*. Unfortunately, the security material is presently so incomplete that it is hard to say what exactly will become a standard (i.e., which algorithms will be recommended); however, it appears clear that some substantial material on IN security (based on the existing security standards) will be part of CS-2. The other open issue is the support of the SCF-to-SCF interface for internetworking. Interestingly enough, the internetworking (independent though of whether it is carried over the SCF-to-SCF or SDF-to-SDF interface) is being discussed in CS-2 in the context of its support of wireless services (i.e., hand-over and profile transfer). On the other hand, it has been decided that the CS-2 protocol for wireless will *not* be defined in CS-2, for which reason its description—based on the IFs of Draft Recommendation Q.1224—has been relegated to an appendix (i.e., the informative rather than normative text) of Draft Recommendation Q.1228. Naturally, until Recommendation Q.1228 is stable, as far as its principles and the outline are concerned, there will be no relevant text in Recommendation Q.1221.

6.2 Recommendation Q.1205, Intelligent Network—Physical Plane

6.2.1 Summary

Recommendation Q.1205 is seven pages long. The first (and still current) publication of the Recommendation is dated March 1993. This is the first Recommendation in the series (and, for that matter, the only one) that defines the IN PEs and assigns to them the FEs of the DFP. The Recommendation also lists the interfaces among the PEs. The definition of the PEs and their relation to FEs is the subject of the following section.

6.2.2 IN Physical Entities and their relation to Functional Entities

Figure 6.1 (after Fig. 1/Q.1205 of the Recommendation) depicts the above entities and all involved interfaces, which are as follows:

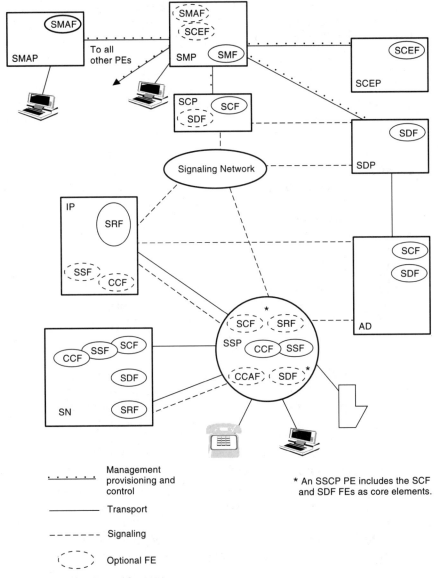

Figure 6.1 (*After Fig. 1/Q.1205.*)

Service Switching Point (SSP), which is a switch that provides access to IN capabilities.[216] An SSP contains a Call Control Function (CCF) and a Service Switching Function (SSF). If it is a local exchange, it may also contain a Call Control Agent Function (CCAF). In addition, an SSP may contain a Specialized Resource Function (SRF). Finally, an SSP may optionally contain a Service Control Function (SCF) and a Service Data Function (SDF), although—one may argue—IN standards are definitely not written with only this type of switch in mind.

Service Control Point (SCP), which contains an SCF and (optionally) an SDF. The SCP has access to the SS7 network, which it uses to communicate with SSPs and IPs. As far as the latter are concerned, the Recommendation, in concert with Recommendation Q.1214 (cf. the SRF interfaces discussion in the previous chapter), declares that an IP may also be accessed via the SSP relay.

Service Switching and Control Point (SSCP), which combines the SCP and SSP in one entity. It contains all the FEs that either PE may contain, but none of them is optional except the SRF. Furthermore, the interface between the SSF/CCF and SCF (internal) FEs is proprietary as is the internal SCF-to-SDF interface; however, all the entities also support the external standard interfaces, and so the SSF/CCF within the SSCP may, for example, send a query to an SCP.

Service Data Point (SDP), which contains only the SDF. The SCP can access data in an SDP either directly or through a signaling network. The key to understanding the role of the SDP is in the statement that it "may be in the same network as the SCP, or in another network." In other words, the SDP is defined as an entity that can be accessed from another network. To this end, it is the only entity that has been *explicitly* specified for internetworking.[217]

Intelligent Peripheral (IP), which contains an SRF and, optionally, an SSF/CCF to provide external access to resources. The Recommendation provides a (nonexhaustive) list of the IP capabilities, which include customized and concatenated[218] voice announcements, synthesized voice, speech recognition, Dual Tone Multi-

---

[216]Note that providing access to IN capabilities is a complex implementation problem, and it should not be taken for granted that any switch may be an IN switch [cf. the description of a Network Access Point (NAP) in the next section].

[217]An important nuance here (recall the discussion of internetworking in the previous chapter) is that it is the logical interface from the SCF to SDF that is sanctioned to cross the networks. For this reason, an SCP in one network *may* communicate with an SCP in another network as long as the SCF-to-SDF part of the protocol is used.

[218]The word "concatenated" refers to a capability to mix and match the announcements sent in one request to the IP.

Frequency (DTMF) digit collection, audio-conference bridging, tone generation, texts-to-speech synthesis, as well as protocol conversion.

An *Adjunct* (AD), which is functionally equivalent to an SCP, but is connected to a single switch via a high-speed network[219] rather than via the SS7 network. The Recommendation notes,[220] though, that "the application layer messages [used at the SCP-to-AD interface] are identical in content[s] to those carried by the signalling network...[at the SCP-to-SSP interface]," which means that there is no difference between the AD and SCP as far as the protocol is concerned.

A *Service Node* (SN), which is similar to an AD, but in addition to performing a role of an SCP, it can also perform that of an IP. The SN may communicate with more than one SSP, but it must have a direct point-to-point signaling and transport connection[221] to each SSP it is communicating with, and each of them must have a point-to-point signaling and transport connection. As far as FEs go, the SN contains the SCF, SDF, SRF, and SSF/CCF. As for the latter, the Recommendation points out (in Clause 3) that "this SSF/CCF is closely coupled to the SCF within the SN, and is not accessible by external SCFs."

A *Service Management Point* (SMP), through which service management and service provisioning are performed. The SMP contains the Service Management Function (SMF), and it may contain a Service Management Access Function (SMAF)[222] and a Service Creation Environment Function (SCEF). The SMP is connected to all PEs (although not through the SS No. 7 network), as is expected because of its provisioning responsibility.

A *Service Management Agent Point* (SMAP),[223] which pretty much serves as a terminal concentrator. (The Recommendation also mentions its possible function as the means of accessing different SMPs from one terminal.) It supports interfaces to computer terminals that

---

[219]Such as a local area network (LAN).

[220]In Clause 3, Item e.

[221]In an implementation of the Service Circuit Node (SCN) built by AT&T Network Systems, this interface is an ISDN channel (either BRI or PRI).

[222]The presence of the SMAF in an SMP means that the latter supports an application interface to computer terminals. It should also be noted that by some (unfortunately inevitable, given the amount of information in the IN Recommendations) glitch in communications between various groups that worked on different Recommendations, the terminology is not entirely consistent through Recommendations. Thus, the SMAF, for example, is called Service Management *Agent* Function in Recommendation Q.1204, but Recommendation Q.1205 calls it Service Management *Access* Function. This point was recently addressed in the IN group, and it has been agreed that the word *Agent* should be used throughout the Recommendations.

[223]Originally, *Service Management Access Point* (cf. footnote 222).

are attached to it, as well as the interface to the SMP. The SMAP contains an SMAF.

A *Service Creation Environment Point* (SCEP), in which the services are programmed and tested. The SCEP contains an SCEF. It is connected to the SMP (although not through the SS No. 7 network).

6.3 Recommendation Q.1208, General Aspects of the Intelligent Network Application Protocol

6.3.1 Overview

Recommendation Q.1208 is only one page long. Its first (and still current) publication is dated March 1993.[224]

The role of this Recommendation is to serve as an introduction to all the protocol specification series (i.e., the Q.12x8 series). At the same time, the Recommendation is the first in the series to state the requirements for the specification of the Intelligent Network Application Protocol (INAP).

For the purpose of this book, we use the Recommendation as a set of reference points that helps us lay the ground for the gradual introduction of the new material. The information presented in the preceding chapters was more or less self-contained and particular to the subject of IN. The material in the remainder of this chapter, however, draws as much from IN as from what had been developed in the many years of collaboration on Signaling System No. 7 (SS No. 7) in ITU, and just about as many years of work on the Open Systems Interconnection (OSI) standards in the International Organization for Standardization (ISO) and ITU.

To deal with this surge of information, we interweave the discussion of the Recommendation with the brief explanation of the terms and standards to which the Recommendation refers.

We review the Recommendation following the taxonomy of Holzmann (1991) introduced in the previous chapter,[225] according to which a protocol specification should include the following five elements:

1. The assumptions about the *environment* where the protocol is executed

---

[224]A new version of the Recommendation is going to be published along with CS-2, but there will be no drastic change in the material. The principles of INAP are going to stay unchanged.

[225]We intentionally changed the order of the elements to suit present discussion.

2. The *service* to be provided by the protocol

3. The *vocabulary* (i.e., the set of protocol messages)

4. The *encoding* of each message

5. The *procedures* (i.e., the rules that sequence the messages)

In the remainder of this section, we address each of these elements in a separate subsection.

6.3.2 INAP environment

Recommendation Q.1208 implicitly specifies the environment as that provided by the Application Layer of the OSI.[226] To this end, INAP provides to the application (and is, in turn, supported by) the Transaction Services.[227] INAP provides its services to the *Application Process* (AP) through what is called *Application Entity* (AE).

An AE may be thought of as an abstract specification of the communication capabilities that an AP can invoke. An AE contains *Application Service Elements* (ASEs),[228] which, according to ITU-T Recommendation Q.775,[229] are "integrated set[s] of actions that ha[ve] a potential to be used in more than one AE." For the purposes of IN Recommendations, ASE can be thought of as a collection of procedure call (also known as *operation*) specifications, such a view being especially appropriate to INAP because, as Recommendation Q.1208 postulates, it is the user of the *Remote Operations Service Element* (ROSE) discussed later in this section. An AP establishes the *association* (i.e., a pipe through which it communicates with another process) and then executes the operations. Generally, this may not be done in an arbitrary fashion; certain rules apply to the order in which the operations

---

[226]ITU-T Recommendation Q.1400, "Architecture Framework for the Development of Signalling and OA&M Protocols Using OSI Concepts," is a good source of comprehensive material on this subject. A recently published monograph [Grinberg (1995)] contains an excellent tutorial on the Application Layer issues related to the specification of Switch-Computer Application Interface (SCAI) protocol, which is in many ways related to the INAP.

[227]Two other types of application services defined in ITU-T Recommendation Q-1400 are *Call Control Application Services* (CCAS) [combining ISUP and Telephony User Part (TUP) protocols] and *Operations, Maintenance and Administration Part* (OMAP) services. As far as the Transaction Services are concerned, they are built over *Transaction Capabilities Application Part* (TCAP) of SS No. 7 supported by the services of the connectionless *Signaling Connection Control Part* (SCCP). The SCCP is in turn supported by the services of the *Message Transfer Part* (MTP) protocol, which combines Layers 1 through 3 of the OSI model. Mitra and Usiskin (1995) provides a comprehensive comparison of the OSI and SS No. 7 protocol stacks.

[228]An AE also contains certain functions that are discussed further in Sec. 6.3.6.

[229]This Recommendation was published in March 1993; the work on its revision continues in ITU-T Study Group 11.

are executed. The rules concerning the order of execution of the operations over one association are enforced by the *Single Association Control Functions* (SACFs). If there are several associations,[230] an additional set of rules that synchronize the communications over these associations may be needed. These rules are enforced by the *Multiple Association Control Function* (MACF). The definitions of the SACF and MACF thus indicate that their combination forms *protocol procedures,* but we discuss those later. The reason we mentioned them here is to complete the description of the structure of an AE. We are about to achieve that with one more definition. Given an association, all the pertinent ASEs together with the SACF form a *Single Association Object* (SAO). The structure of an AE is depicted in Fig. 6.2.

Speaking figuratively, one can view a person making a simple telephone call as an AP and the telephone terminal itself as an AE. This AE contains the following ASEs: the off-hook/on-hook switch (the Switch ASE); the set of buttons that correspond to digits (the Digits ASE); and the set of buttons that correspond to special characters (* and #), etc. In addition there may be a set of special buttons with preprogrammed numbers (the Address ASE). These ASEs are dealing with establishing a call (an association) over the telephone network. The SACF must contain the rule that the telephone must be off-hook before the digits are dialed; the SACF may also contain the rules for forming a correct number, etc. If the telephone has two lines, then the MACF may contain the rules that govern switching from one line to another as well as the rules for joining and splitting the lines. (There may be additional ASE buttons for these functions.)

Now, as promised, we return to the discussion of ROSE. ROSE is a special standardized[231] ASE itself. Because it provides the facilities for a remote procedure call, it is used in many distributed processing applications. Its significance to INAP warrants a closer look at it; its relative simplicity allows us to do so in this book.

ROSE specifies four *Protocol Data Units* (PDUs):[232] *Invoke, Return Result, Return Error,* and *Reject.* To understand the ROSE model, con-

---

[230]As there are in the case of the SCP, which, on behalf of any particular call, may simultaneously deal with three associations: one with the SSF/CCF, one with the SDF, and one with the SRF.

[231]See ITU-T Recommendations X.219 and X.229 (aligned with ISO 9072-1 and ISO 9072-2, respectively), published in 1988, as well as ITU-T Recommendation X.880 (1994) or ISO/IEC 13712-1:1994, Information technology, "Remote Operations: Concepts, model and notation."

[232]In the OSI terminology, a PDU is a message sent from one entity to its peer in the same layer. The term has been used mostly to denote the PDUs in upper layers of the OSI model (Tanenbaum, 1989): Transport Layer PDUs (TPDUs), Session Layer PDUs (SPDUs), and Application Layer PDUs (APDUs). Naturally, all PDUs relevant to this book are APDUs.

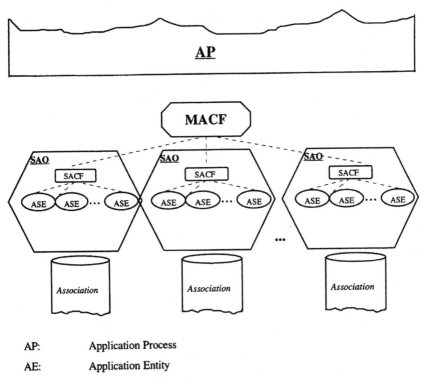

AP:	Application Process
AE:	Application Entity
ASE:	Application Service Element
SACF:	Single Association Control Function
SAO:	Single Association Object
MACF:	Multiple Association Control Function

Figure 6.2 Application Entity structure.

sider two AEs that communicate with each other. One of these AEs is the user (client), and the other is the provider (server). The user issues its request for an *operation* to be performed by a provider (the semantics of the operation being transparent to ROSE) via a procedure call corresponding to the ROSE Invoke PDU.[233] This *operation* is then carried— enveloped within the *Invoke* PDU—to the server. The provider attempts to perform the operation, at which point it may send back its result by using the *Return Result* PDU. If the operation fails, the provider may send back the error message conveyed within the *Return Error* PDU.

It is also possible that the server does not recognize the request, in which case it may send back the *Reject* PDU. Each operation request explicitly specifies whether the response is expected back, and, if so, in

---

[233]The parameters of the procedure call contain the operation-related information.

what manner it should be sent. To this end, four classes of operations are specified:[234]

- Class 1 operations require that both success and failure be reported.
- Class 2 operations require that only success be reported.
- Class 3 operations require that only failure be reported.
- Class 4 operations require no report at all.[235]

Finally, we introduce one last concept related to the Application Protocol environment, which is necessary for understanding INAP. The name of the concept is *Application Context*. This time we precede the definition with an example. Going back to the person-AP in front of the telephone-AE, we find this person in the act of starting a conversation. We consider the words and even whole sentences the *operations*. Those are grouped into (rather huge) topical ASEs (the food ASE, the music ASE, etc.). Still at the beginning of the conversation, both interlocutors are presently agreeing on the subject (e.g., a five-course dinner, based on the food ASE and governed by SACF rule that the courses are discussed in the order they are to be consumed: appetizer, soup, meat or fish, cheeses, and dessert). What is being agreed on here is the application context.

Note that when we spoke of the words and sentences, we did not mention the language in which the conversation was taking place. Actually, both gluttons just happened to speak English, but one of them could speak only Russian and English and the other, only French and English. In this case, they would soon discover that they were not speaking the same language and would negotiate the common language. In this negotiation they would determine the *Presentation Context,* of which the application context is invariant.[236] (The images of food in one's head are independent of the language one speaks.)

---

[234]The OSI ROSE actually has five classes of operations, but, for the purposes of TCAP protocol, only four (OSI ROSE classes 2–5, respectively) had been selected.

[235]Note that the first three classes of operations involve the user's waiting for a response. If the ROSE primitives are implemented as procedure calls that return to the calling process when the operation has either completed or failed, an elaborate scheme of timers needs to be maintained on both sides to ensure that the calling process does not hang. We attract the reader's attention to this issue because it will arise when we discuss the INAP.

[236]Just before this sentence was written, one author had to call a hotel in Miyazaki, Japan (where a Study Group 11 meeting was planned) to make a reservation. After sheepishly inquiring whether the woman who answered the call spoke English and being answered in the affirmative (successful negotiation of the presentation context), the author mentioned that he wished to make reservations (i.e., suggested an application context). At this point, the communications faltered momentarily, but then the call was transferred to the reservation desk (another association was established), where the author's application context was accepted.

This should make the concept of application context intuitively clear. Formally, it may be defined as a set of ASEs and rules that are to govern the communications among APs. An AP that initiates the communications presents one or more contexts in a PDU and receives a response, which either confirms the agreement to use the proposed context, denies it, or, proposes another context. For the latter option to take place, the association (or transaction) has to be closed—and the new one started—in order to propose another set of contexts.

6.3.3 INAP service

The semantics of the services provided by INAP is specified at the DFP. The role of INAP is to carry the information defined in the IFs and their IEs exchanged between the FEs. Why FEs and not PEs? Clause 3 of Recommendation Q.1208 stresses that "the protocols should be defined in such a way that the Functional Entities defined...may be mapped into Physical Entities in any way that operators and manufacturers desire."

6.3.4 INAP vocabulary

The *vocabulary* of the INAP consists of the ROSE-supported *operations* and their *parameters,* which respectively correspond to their DFP counterparts: the IFs and IEs. By now the reader should have developed enough understanding of the INAP environment to see how *in principle* the IFs and IEs of the DFP can be used in the environment described above. Since it is Recommendation Q.1218 that contains the mapping of the IFs and IEs into the operations and parameters, we defer the respective discussion until Sec. 6.6.

6.3.5 INAP encoding

It used to be in data communications until very recently that the structure of each message (i.e., PDU) was specified by providing its precise bit-by-bit layout. There was merit to this approach because it was unambiguous, as far as the implementations were concerned, and it was efficient (about as many bits were sent in a message as were needed). The latter quality—still desirable even in the present high-bandwidth era—was decisively important even a decade ago.

This approach had one serious disadvantage, however: the human factor. Even though the protocols were invented for computer communications, it was the humans who were to design, program, and test these protocols. Aside from the unpleasant feelings naturally brought by the act of staring at the strings of 0's and 1's, consider what one had to go through when switching from one field (i.e., a part of the string that corresponds to an individual parameter) to another. For example, a field that corresponded to an integer number (encoded as, say, "00010010," denoting 18) could be followed by a field that corresponded to a power-set[237] of elements (where an encoding "101" might mean that green and red colors are to be applied but the blue is not). While the encoding of "infinite" structures (such as lists) has always given programmers headaches, even simple atomic data, such as integer numbers, could be troublesome. One had to be especially careful with the negative numbers because those can be encoded in at least three different (and commonly used) ways. Even encoding (and decoding) of positive numbers presents problems to a human because different computer architectures often use different bit layouts. In other words, the least significant bit can be either the left-most or right-most one in an *octet* (or *byte,* in computer parlance); the same holds regarding the layout of bytes in a 16-bit *word,* 16-bit words in 32-bit words, etc. Another serious problem is that the protocol messages are often standardized to contain optional fields, that is, the fields that, depending on the implementation (or even the run-time circumstances) of the protocol, may or may not be present.[238] Obviously, by glancing at the bit string neither a computer nor a human could, in general, be sure which fields were present in any particular message. The practical remedy in such cases is to tag the fields (i.e., assign them names,[239] which then must always accompany the fields that are optional), which leaves the persons responsible for debugging of the protocol implementations with the additional work of decoding the names of the fields *before* the grisly task of decoding the fields themselves begins. Naturally, these types of problems significantly limit the productivity of protocol designers and implementers.

Incidentally, the last few examples, which mention such constructs as power-sets, lists, and optional parameters, have not been *invented* to demonstrate how bad the old approach was; these examples actually refer to the language constructs, which have been present in all modern general-purpose high-level languages (e.g., *ALGOL, Pascal, C, Common LISP,* or *C++*),s as well as the database query languages (e.g., *SQL*), in which the majority of the applications are written. It is intrin-

---

[237]A *power-set* is an abstract data structure that supports the representation of and operations on the finite sets of elements. It has been traditionally implemented as a bit string corresponding to the characteristic function of the set: one bit of this string is assigned to each element; the value 1 of a bit signifies the presence of the element in the set, and the value 0, its absence.

[238]In INAP, for example, *most* of the parameters are optional. This, by the way, by no means spells a deficiency of the standard. On the contrary, it means that the standard is flexible enough to cover many particular cases.

[239]As a rule, integer numbers have been used for this purpose.

sic then that the Application Layer PDUs must carry the data in a way that reflects their use in programs.

Addressing the problem of encoding, we note that the history of data communications repeated the history of computing. As in the latter, it was recognized that (1) the computer programs can be written in a high-level language whose syntax is much closer to the natural human languages than the sequences of 0's and 1's of the machine code—the multitude of different machine codes, for that matter—and (2) a high-level language can be automatically translated into a machine code, so in the former it was recognized that (a) the *abstract syntax* of a message, which should be the only one known to the protocol designer, can be expressed by a high-level language and (b) the *transfer syntax* (i.e., the layout of the actual bit string sent across the network) can be automatically derived from the abstract syntax.

Thus the new language *Abstract Syntax Notation One (ASN.1)* was defined and standardized.[240] Recommendation Q.1208 specifically prescribes the use of this language[241] for specification of INAP.

ASN.1 is a Pascal-like language for the encoding-independent definition of Application Layer PDUs (which, as far as the computer languages are concerned, are data structures). ASN.1 contains the set of atomic data types (*INTEGER, REAL, BOOLEAN, ENUMERATED, BIT STRING,* and *OCTET STRING*) as well as the constructs for forming structured data types using these atomic types.

---

[240]The original ASN.1 standard that prescribes the MACRO notation used in INAP is contained in CCITT Recommendation X.208 I ISO/IEC 8824: 1987, Information Processing Systems—Open Systems Interconnection—"Specification of Abstract Syntax Notation One (ASN.1)." [See also Mitra I (1995).] The subsequent standard, also relevant to INAP (specifically to the part that specifies the SCF-to-SDF protocol over X.500) is contained in the following ITU-T Recommendations: Recommendation X.680 (1994) or ISO/IEC 8824-1:1994, Information technology-Open Systems Interconnection—"Abstract Syntax Notation One (ASN.1): Specification of basic notation;" Recommendation X.681 (1994) or ISO/IEC 8824-2:1994, Information technology—Open Systems Interconnection—"Abstract Syntax Notation One (ASN.1): Information object specification;" Recommendation X.682 (1994) or ISO/IEC 8824-3:1994, Information technology—Open Systems Interconnection—"Abstract Syntax Notation One (ASN.1): Constraint specification;" and Recommendation X.683 (1994) or ISO/IEC 8824-4:1994, Information technology—Open Systems Interconnection—"Abstract Syntax Notation One (ASN.1): Parameterization of ASN.1 specifications." Mitra I (1995) and Mitra II (1995) are excellent tutorials on both ASN.1 and the encoding rules. Tanenbaum (1989) and Grinberg (1995) also contain insightful examples. ITU-T Recommendation X.690 (1994) or ISO/IEC 8825-1:1994, Information technology—Open Systems Interconnection—"Specification of ASN.1 encoding rules: Basic, Canonical, and Distinguished Encoding Rules."

[241]And, specifically, its MACRO facility. The introduction of the Information Object Class construct took place later, and it was adopted for the SCF-to-SDF interface in the process of CS-1 Refinement, and therefore is not reflected in Recommendation Q.1208. Mitra I (1995) describes both the MACRO facility and the Information Object Class constructs.

We briefly introduce the structured type constructs that are used in INAP:

- The *SET* construct denotes a *structure* (it corresponds to the *RECORD* construct of Pascal, not the *SET OF* one!).

- A *SEQUENCE* is a *SET* whose components are ordered (and therefore *must* be sent in the order they appear in the specification).[242]

- Either type may be used in defining arrays with the help of constructs *SIZE* and *OF*. (For instance, *SEQUENCE SIZE [1..10] OF <type>*.)

- The *CHOICE* construct (which corresponds to that of *CASE* in Pascal) is used to declare alternatives. For example, the *FilteringTimeOut* parameter (which corresponds to the *Filtering Timeout* IE)[243] should specify either the duration of filtering or the time it should stop. This requirement has resulted in the following specification:

FILTERINGTIMEOUT :: = CHOICE {DURATION [0] DURATION,
                              STOPTIME [1] DATEANDTIME}

The first string in each of the two clauses is the identifier; it is followed by its tag[244] enclosed in the brackets; the third (and last) string in the clause denotes the type of the identifier.

- *ANY* denotes any defined type. This construct is extremely useful in specifying generic constructs. To avoid obvious ambiguities, *ANY* should be followed by the *DEFINED BY* clause, which points to the definition of the type.

Again, the advantage of using ASN.1 versus performing detailed encoding is similar to the advantage of writing programs in a high-level language versus producing binary code. Nevertheless, the use of ASN.1 (or rather its outcome) has been sometimes frowned upon in the industry, the reason being that until recently there has been only one standard method of translating ASN.1 specifications into ultimate bit patterns. This method is given by *Basic Encoding Rules* (BER), which

---

[242]INAP uses only the *SEQUENCE* construct.

[243]See the description of the *Limit* SIB in the previous chapter.

[244]Every identifier in ASN.1 is assigned a tag value by which it is recognized at the time of decoding.

results in less-compact messages and thus may affect performance. It is important to remember, however, that future IN Recommendations may permit encoding rules other than BER, such as Packed Encoding Rules (PER).[245]

6.3.6 INAP procedures

Protocol procedures synchronize the activities related to both receiving and sending messages between the communicating entities. The procedures form the crux of the distributed processing. While the errors with the syntax of the protocol (i.e., missing parameters) can be easily found and corrected by humans, the issues of synchronization are so complex as to prove virtually intractable at the design stage.[246] Furthermore, procedure errors often result in such bizarre behavior of the system that it is extremely hard to determine their causes. And to make the matters worse, such incidents may be intermittent and impossible to recreate.

In ITU-T Recommendations, the protocol procedures have been usually specified by the combination of the "call flows" (i.e., specific exchanges among the communicating entities required to demonstrate the support of a given capability) and the SDL descriptions (one for each communicating entity). The call flows are instrumental in illustrating the main idea behind the protocol, but they can only be illustrative because they cannot capture the infinite aggregations of the message exchanges required to cover the error cases. The SDL descriptions, on the other hand, can perfectly capture all the cases; furthermore, special tools can be used to verify the SDL descriptions. The disadvantage of SDL descriptions is that they often tend to spread over many pages and thus are rather hard to grasp. During the initial work on the IN standard, a pithy BCSM-like FSM-based description of the IN entities did the job. Thus, Clause 3 of Recommendation Q.1208 states that the procedures are defined "in terms of state transition diagrams" and that "other definition techniques may be used in future capability sets to supplement or replace these if this seems appropriate."[247]

---

[245]The relevant standards are the ITU-T Recommendations X.690 (1994) I ISO/IEC 8825-1:1994, Information technology—Open Systems Interconnection—*Specification of ASN.1 encoding rules: Basic, Canonical, and Distinguished Encoding Rules* and X.691 (1995) I ISO/IEC 8825-2: 1995, Information technology—Open Systems Interconnection—*Specification of ASN.1 encoding rules: Packed Encoding Rules (PER)*. Mitra II (1995) provides an in-depth treatment of the subject.

[246]There are formal methods in existence today, and they have been used in standards (see Ruggles, 1990). SDL, discussed before, is also used in formal specification of systems.

[247]The refined version of CS-1 already contains (in an annex to Recommendation Q.1218) the full set of the SDL descriptions of the procedures.

6.4   Relevant Aspects of Recommendation Q.1211, Introduction to Intelligent Network Capability Set 1

6.4.1   Protocol-related aspects

Recommendation Q.1211 reiterates[248] the requirement of Recommendation Q.1208 that the INAP be defined on the FE level rather than the PE level so as to facilitate "flexible packaging of the SSF, SCF, SDF, and SRF into a variety of different Physical Entities."

Recommendation Q.1211 further postulates that the IN Application Service Elements (ASEs) in CS-1 are to be aligned with the ISO standard IS 9545 and be defined independent of the protocol stacks. The recommended protocol stacks are SS No. 7 TCAP and DSS 1 over Q.932.[249] Each of the IN-specific interfaces [i.e., SCF-to-SSF (D), SCF-to-SRF (E), and SCF-to-SDF (R)] is to have a separate set of ASEs. Furthermore, the Recommendation defines three groups of CS-1 ASEs[250] in order to reflect "the capabilities that can be differentially applied to separate classes of services." Before introducing these groups, we include the following historical note.

Recommendation Q.1211 had been completed—by design—before the rest of the CS-1 Recommendations. After all, the purpose of the Recommendation was to define the scope of CS-1. Historically, at the time the Recommendation was to be "frozen," it was still unclear whether the call party handling capabilities would be included in CS-1. In cases like that, the usual procedure in standards is to prioritize the work and then schedule it so as to ensure that the high-priority items are complete. The grouping of ASEs, in addition to serving the immediate purpose of their introduction explained in the previous paragraph, also provided a natural mechanism for prioritization of the capabilities.

Three groups of ASEs were thus defined: (1) the group that supports just the capabilities necessary for call setup, (2) the group that supports midcall control, and (3) the group that supports "call topology manipulation."[251] Of these, only the first group (which, by the way,

---

[248]In Clause 7.5.

[249]Because Recommendation Q.1211 was not reissued as the result of the CS-1 refinement process, it does not explicitly specify the X.500 (directory) protocol (brought into IN during the refinement process), on which the SCF-to-SDF interface is presently based; however, regarding the SDF-to-SDF and SMF-to-SMF interfaces, Clause 7.7 of the Recommendation states that "wherever possible, the CS-1 Recommendations identify alternative interfaces (e.g., SCCP-GTT, X.500, or CMISE) for these domains."

[250]In Clause 7.5.

[251]This beautiful sounding name replaced the suspicious "leg manipulations" (so-called because of the names of operations used for this capability: "SplitLeg," "AttachLeg," etc.) but was later replaced itself with "Call Party Handling."

includes routing, charging, and user interaction capabilities) was declared to be of high priority.

The Recommendation does not define the ASEs themselves. This is done in Recommendation Q.1218.

6.5   Recommendation Q.1215, Intelligent Network—Physical Plane for CS-1

6.5.1   Summary

Recommendation Q.1215 was initially published by ITU-T in March 1993, at which time it had nine pages. The next (CS-1 Refined) version of this document was approved in May 1995. The material of this section is based on the final draft text (Faynberg, 1995 II), which does not differ in size from the original one.

Recommendation Q.1215 excludes the service creation and service management PEs (SCEP, SMAP, and SMP) and the relevant interfaces from the CS-1 architecture, but it introduces an additional (to those defined in Recommendation Q.1205) PE, called *Network Access Point* (NAP) and changes options in the mapping of some FEs to PEs. Finally, the Recommendation specifies the protocol to be used for all CS-1 PE-to-PE interfaces.

The following two sections describe the CS-1 entities and the protocol selection.

6.5.2   CS-1 entities

The CS-1 entities (see Fig. 6.3, drawn after Fig. 1/Q.1215 of the Recommendation) are as follows: Service Switching Point (SSP), Service Control Point (SCP), Service Data Point (SDP), Service Switching and Control Point (SSCP), Adjunct (AD), Intelligent Peripheral (IP), Service Node (SN), and a new (in respect to Recommendation Q.1205) PE, *Network Access Point* (NAP).

A NAP is a switch that includes only the CCAF and CCF (but not the SSF) FEs. As the Recommendation explains, "the NAP supports early and ubiquitous deployment of IN-based services. This NAP cannot communicate with an SCF [because a NAP does not have the SSF], but it has the ability to determine when IN processing is required. It must send calls requiring IN processing to an SSP."[252]

---

[252]The necessity of introducing the NAP was dictated by the presence of admirable and not fully appreciated economically pre-IN switches (such as the AT&T NS 1A ESS of the Regional Bell Operating Companies in the United States). These switches can still be fully used in the network for non-IN-supported calls, and they can be relatively inexpensively upgraded to route the IN-supported calls to the SSP switches.

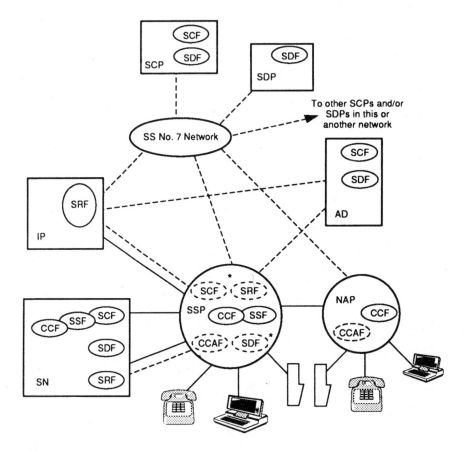

* An SSCP PE includes the SCF and SDF FEs as core elements

——————— Transport

- - - - - - Signaling

(⸜⸝) Optional FE

Physical entities (PEs)
AD   Adjunct
IP   Intelligent peripheral
SSP   Service switching point
SCP   Service control point
SN   Service node
NAP   Network access point
SDP   Service data point
SSCP   Service switching and control point
        Functional entities (FEs)
CCF   Call control function
CCAF   Call control agent function
SCF   Service control function
SDF   Service data function
SRF   Special resource function
SSF   Service switching function

Figure 6.3 Physical Plane architecture for CS-1. (*After Fig. 1/Q.1215.*)

There are two main differences with Recommendation Q.1205, as far as the mapping of the FEs to PEs is concerned. First, Recommendation Q.1215 explicitly states[253] that the SCP *must* contain *both* the SCF and the SDF, while the latter FE is declared optional[254] in Recommendation Q.1205. Secondly, Recommendation Q.1215 assigns[255] only the SRF to IP, while Recommendation Q.1205 allows[256] the IP to contain the SSF/CCF as an optional FE.

6.5.3 Protocol selection

Recommendation Q.1215 proposes the Transaction Capabilities Application Part (TCAP) for the SCP-to-SSP, SCP-to-IP, SCP-to-SDP (when both the SCP and SDP are in the same network),[257] AD-to-SSP, and AD-to-IP interfaces. In addition, for the former three interfaces, it states that TCAP is to be used over the SS No. 7 Signaling Connection Control Part (SCCP)/Message Transfer Part (MTP) protocol platform.

For the SSP-to-IP interface, the Recommendation proposes the ISDN *Basic Rate Interface* (BRI) or *Primary Rate Interface* (PRI) (or both) as well as the SS No. 7. The Recommendation clarifies that if the ISDN interfaces are used, the D-channel connecting an IP to an SSP is to be used for relaying the application layer information between an SCF and an SRF as well as for setup of B-channel connections to the IP. The Recommendation also says that the IEs in the SCF-to-SRF IFs are to be embedded in the Facility Information IEs (which are present in Q.931 messages). A similar arrangement (i.e., the ISDN D-channel as the transport) is suggested by the Recommendation for the SN-to-SSP interface, with reference to the procedures of ITU-T Recommendation Q.932.

Finally, the Recommendation lists the options[258] for the end-user-to-network access interfaces by stressing the requirement that the existing network interfaces be supported in IN deployment. The existing interfaces define what information is available to IN. To illustrate this point, the Recommendation cites an example of the calling party information, which "may or may not be available at a given interface and, therefore, may or may not be provided to the SCF." Both, the analogue interface signaling and ISDN access signaling arrangements are to be supported in CS-1.

---

[253]In Clause 5.1 of Recommendation Q.1215.

[254]In Clause 4.1 of Recommendation Q.1205.

[255]In Clause 5.1 of Recommendation Q.1215.

[256]In Clause 4.1 of Recommendation Q.1205.

[257]If they are in different networks with incompatible protocols, the Recommendation proposes the use of protocol converters.

[258]In Clause 5.3.8.

6.6   Recommendation Q.1218, Intelligent Network Interfaces

6.6.1   Summary

Recommendation Q.1218 was initially published by ITU-T in March 1993, at which time it had 110 pages. The next (CS-1 Refined) version of this documentation was approved in May 1995. The material of this section is based on the final draft text (Turner, 1995), which has about 500 pages.

The CS-1 refinement work has affected this Recommendation the most. For this reason, the reader may find the following historical material interesting. The first refinement process was started by the European Telecommunication Standards Institute (ETSI) as part of the work on the European IN standard. The process resulted in new ETSI Recommendations, of which the counterpart of Recommendation Q.1218 is the ETSI Core INAP Recommendation.[259] By far the most dramatic development of this activity, as far as the protocol is concerned, was the removal by ETSI of all DP-specific (and some other) operations.[260] For the remaining operations and their parameters, ETSI provided precise description of their purpose and use. ETSI also enhanced the procedure model. Finally, ETSI brought its newly created standard to ITU as a contribution to the refinement process.

The resulting refined CS-1 INAP retained all operations, but augmented their description, according to the ETSI-proposed methodology; the procedural model was also improved based on the ETSI work. In addition, the refined CS-1 INAP adopted the Directory standard for the SCF-to-SDF interface.

The final version of the Recommendation contains

1. Detailed background information on the protocol stacks supporting the INAP and the relevant standards (both ITU-T and ISO).

2. The full vocabulary of the INAP (i.e., operations and parameters) expressed in ASN.1. The description of the operations is combined with the description of the IN timers, which adds a procedural overtone to the vocabulary specification.

3. The specification of the INAP ASEs.

4. The specification of INAP *Application Contexts* (ACs).

---

[259]European Telecommunications Standard ETS 300 374-1, Intelligent Network (IN); Intelligent Network Capability Set 1 (CS-1) Core Intelligent Network Application Protocol (INAP) Part 1: Protocol specification. Sophia Antipolis, France: European Telecommunications Standards Institute, September 1994.

[260]Rinker (1995) provides a detailed comparison of ETSI Core INAP and CS-1-refined INAP.

5. The model of the distributed IN environment, and the FSM-based description of the IN procedures.[261]

6. The detailed description of the purpose and use of all operations and their parameters. This material is also referred to as procedures, although it complements the material mentioned in the previous list item, which focuses on sequencing of the operations and ignores the parameters altogether.

Recommendation Q.1218 also has an annex, which contains the SDL-based implementation of the procedural model, and two appendices: Appendix I outlines the guidelines for the definition of the service data model (based on the information model of Recommendation X.501), and Appendix II is the repository of the "for further study" protocol material (related to the Call Party Handling capabilities).

The rest of this section reviews the vocabulary of INAP (including specification of the ASEs and ACs) and its procedures.

6.6.2 The INAP vocabulary

The INAP vocabulary contains the definitions of operations, parameters,[262] ASEs, and ACs described in the sections following.

6.6.2.1 Operations. The specification of the operations used at the SSF/CCF-to-SCF and SRF-to-SCF interfaces[263] differs from that of the operations used at the SCF-to-SDF interface.[264] The former operations are carried directly by ROSE-based TCAP, and their specification uses the *MACRO* notation. The latter operations are carried via the Directory protocol,[265] and their specification uses the Information Object construct of the later version of ASN.1. With either mechanism, however, the specification of a particular operation includes

- The *ARGUMENT* clause, which references the parameters of the operation

---

[261]Effectively, these are the SACF and MACF rules.

[262]The parameters are not addressed in this book; a reader who wishes to get full information on the subject should consult Clause 3.3 of the Recommendation, which provides a self-contained and straightforward description of the parameters and their use in the operations on an operation-by-operation basis.

[263]In Clause 2.1 of the Recommendation.

[264]In Clause 2.2.

[265]The following standards are relevant: ITU-T Recommendation X.500 (1993) or ISO/IEC 9594-1:1993, Information Technology—Open Systems Interconnection—"The Directory: Overview of Concepts, Models and Services," ITU-T Recommendation X.501 (1993) or ISO/IEC 9594-2:1993, Information Technology—Open Systems Interconnection—"The Directory: The Models," and ITU-T Recommendation X.511 (1993) or ISO/IEC 9594-3:1993, Information Technology—Open Systems Interconnection—"The Directory: Abstract Service Definition."

- The *RESULT* clause (present only if the operation is of either class 1 or 2), which references the parameters returned with the result

- The *ERRORS* clause (present only if the operation is of either class 1 or 3), which references parameters returned in error cases

- The *LINKED* clause (optional), which references the operations that may be issued from the other end.[266]

The class of a particular operation is implicit: if both the *RESULT* and *ERROR* clauses are specified, the operation is of class 1; if only the *RESULT* clause is specified, the operation is of class 2; if only the *ERRORS* clause is specified, the operation is of class 3; and, if neither the *RESULT* nor *ERROR* nor *LINKED* clause is specified, the operation is of class 4.

There is a nuance in the specification of DP-specific operations. Recalling the material of the previous chapter, each such operation can be used in INAP either as a request[267]—if it is invoked at a TDP, or a report—if it is invoked at an EDP. Technically, the former could be expected to be a class 1 or class 2 operation,[268] while the latter should be a class 4 operation. All DP-specific operations have been specified as class 2 operations, however. There is nothing drastically wrong with that, but an implementer ought to be aware of this situation.[269]

The operations, errors, and data types for the SSF/CCF-to-SCF interface are listed in separate ASN.1 modules, whose names are *IN-CS-1-Operations, IN-CS-1-Errors,* and *IN-CS-1-Data-Types,* respectively. In addition, the module *IN-CS-1-Codes* specifies the INAP operation and error codes.

Each operation is also assigned a timer, whose value can be *short* (up to 10 seconds), *medium* (up to 60 seconds), or *long* (up to 30 minutes), with the minimum value of 1 second for each of the three categories of timers. For some timers, their categories are not specified by the standard and, for this reason, are marked "For Further Study (FFS)." Although neither the *OPERATION* macro nor the *OPERATION* class of the new ASN.1 has a place for the timer specification, Clause 2 of the Recommendation lists the name of an appropriate timer for each operation, as an ASN.1 comment. The types of the timers are listed in a separate table.[270]

---

[266]For example, the *PlayAnnouncement* operation invoked by the SCF lists the *SpecializedResourceReport* operation invoked by the SRF in the *LINKED* clause.

[267]Similarly to the InitialDP operation.

[268]InitialDP is a class 2 operation.

[269]An SCP implementer may distinguish the reports from the requests and decide never to send error messages to the former.

[270]Table 2/Q.1218.

The reader should have no problem finding the detailed information in the Recommendation. For a quick summary, Table 6.1 contains a mapping of the CS-1 IFs into the INAP operations. Note that a *pair* of IFs that corresponds to the exchange of the request and result (such as the *Authenticate/Authenticate Result* pair) maps to only *one* operation. Finally, two operations,[271] *Bind* and *Unbind* are particular to the Directory protocol used at the SCF-to-SDF interface, which is why they have no IN IFs associated with them.

6.6.2.2 ASEs. Again, as far as Recommendation Q.1218 is concerned, all ASEs are just groups of operations. The operations are presently grouped into ASEs according to a specific function (e.g., traffic management).[272] As in the case of the INAP operations, the description of the ASEs containing the operations at the SSF/CCF-to-SCF and SRF-to-SCF interfaces differs from the description of the ASEs related to the SCF-to-SDF interface. The former ASEs are specified using the *MACRO* notation. The latter ASEs are specified using the Information Object construct of the later version of ASN.1.

The *APPLICATION-SERVICE-ELEMENT* macro contains two clauses, *CONSUMER-INVOKES* and *SUPPLIER-INVOKES,* which are respectively related to the consumer and supplier of the application service. Each clause lists the operations invoked by the entity to which it refers. Again, the ASEs specified in this way are related to the SSF/CCF-to-SCF interface; they are specified in Clause 2.1.4 of the Recommendation.

For the SCF-to-SDF interface, the ASEs are specified (as information objects) in Clause 2.2 of the Recommendation. Two classes are used for this purpose: the *CONNECTION-PACKAGE* class (for the *dapConnectionPackage* information object, which contains the *Bind* and *Unbind* operations) and the *OPERATION-PACKAGE* class (for the *searchPackage* and *modifyPackage* information objects, which contain the remaining operations). In addition, the *dapContract* information object of class *CONTRACT*[273] is specified.

Table 6.2 lists all CS-1 INAP ASEs together with the operations they contain. The names of the SCF-to-SDF ASEs are invented to keep the

---

[271]These operations are used to establish the association between the *Directory User Agent* (DUA) residing in the SCF and the *Directory System Agent* (DSA) residing in the SDF.

[272]The present grouping is very logical in some cases (e.g., the Basic-BCP-DP ASE) and somewhat arbitrary in others. In the future, the ASEs may be regrouped according to SIBs so that the ASE that corresponds to a SIB contains all the operations necessary for the execution of this SIB (and no other operations).

[273]For the momentary purposes of this section only, a "contract" can be viewed as an Application Context. It refers to the connection package and the rest of the packages used by the consumer.

TABLE 6.1   Mapping of Information Flows of Recommendation Q.1214 to Operations of Recommendation Q.1218

Information flows	Operations
Activate Service Filtering	*ActivateServiceFiltering*
Activity Test/Activity Test Response	*ActivityTest*
Add Entry/Add Entry Result	*AddEntry*
Analyze Information	*AnalyzeInformation, Connect*
Analyzed Information	*AnalyzedInformation* (if DP-specific operations are supported); otherwise, *InitialDP* or *EventReportBCSM,* depending on whether it is issued at a TDP or EDP.
Apply Charging	*ApplyCharging*
Apply Charging Report	*ApplyChargingReport*
Assist Request Instructions	*AssistRequestInstructions*
Assist Request Instructions from SRF	*AssistRequestInstructions*
Authenticate/Authenticate Result	*Authenticate*
Call Gap	*CallGap*
Call Information Report	*CallInformationReport*
Call Information Request	*CallInformationRequest*
Cancel Announcement	*Cancel*
Cancel Status Report Request	*CancelStatusReportRequest*
Collect Information	*CollectInformation*
Collected Information	*CollectedInformation*
Connect	*Connect*
Connect to Resource	*ConnectToResource*
Continue	*Continue* (Note: this is not TCAP *Continue* operation, but an INAP one!)
Disconnect Forward Connection	*DisconnectForwardConnection*
Establish Temporary Connection	*EstablishTemporaryConnection*
Event Notification Charging	*EventNotificationCharging*
Event Report BCSM	*EventReportBCSM*
Furnish Charging Information	*FurnishChargingInformation*
Hold Call in Network	*HoldCallInNetwork,* if this operation is supported (it is not in ETSI Core INAP); otherwise, *ResetTimer* or *FurnishChargingInformation,* whatever is to be issued first in a given implementation.
Initial DP	*InitialDP*
Initiate Call Attempt	*InitiateCallAttempt*

TABLE 6.1 Mapping of Information Flows of Recommendation Q.1214 to Operations of Recommendation Q.1218 (Continued)

Information flows	Operations
Modify Entry/Modify Entry Result	*ModifyEntry*
O-Answer	*OAnswer* (if DP-specific operations are supported); otherwise, *InitialDP* or *EventReportBCSM*, depending on whether it is issued at a TDP or EDP.
O_Called_Party_Busy	*OCalledPartyBusy* (if DP-specific operations are supported); otherwise, *InitialDP* or *EventReportBCSM*, depending on whether it is issued at a TDP or EDP.
O_Disconnect	*ODisconnect* (if DP-specific operations are supported); otherwise, *InitialDP* or *EventReportBCSM*, depending on whether it is issued at a TDP or EDP.
O_Mid_Call	*OMidCall* (if DP-specific operations are supported); otherwise *InitialDP* or *EventReportBCSM*, depending on whether it is issued at a TDP or EDP.
O_No_Answer	*ONoAnswer* (if DP-specific operations are supported); otherwise, *InitialDP* or *EventReportBCSM*, depending on whether it is issued at a TDP or EDP.
Origination Attempt Authorized	*OriginationAttemptAuthorized* (if DP-specific operations are supported); otherwise, *InitialDP*.
Play Announcement	*PlayAnnouncement*
Prompt and Collect User Information/Collected User Information	*PromptandCollectUserInformation*
Release Call	*ReleaseCall*
Remove Entry/Remove Entry Result	*RemoveEntry*
Request Notification Charging Event	*RequestNotificationChargingEvent*
Request Report BCSM Event	*RequestReportBCSMEvent*
Request Status Report	*RequestCurrentStatusReport,* or *RequestFirstStatusMatchReport*, or *RequestEveryStatusChangeReport*, depending on the application needs.
Reset Timer	*ResetTimer*
Route Select Failure	*RouteSelectFailure* (if DP-specific operations are supported); otherwise, *InitialDP* or *EventReportBCSM*, depending on whether it is issued at a TDP or EDP.
Search	*Search*

TABLE 6.1   Mapping of Information Flows of Recommendation Q.1214 to
Operations of Recommendation Q.1218 (*Continued*)

Information flows	Operations
Search Result	*SearchResult*
Select Facility	*SelectFacility*
Select Route	*SelectRoute*
Send Charging Information	*SendChargingInformation*
Service Filtering Response	*ServiceFilteringResponse*
Specialized Resource Report	*SpecializedResourceReport*
Status Report	*StatusReport*
T_Answer	*TAnswer* (if DP-specific operations are supported); otherwise, *InitialDP* or *EventReportBCSM*, depending on whether it is issued at a TDP or EDP.
T_Called_Party_Busy	*TBusy* (if DP-specific operations are supported); otherwise, *InitialDP* or *EventReportBCSM*, depending on whether it is issued at a TDP or EDP.
T_Disconnect	*TDisconnect* (if DP-specific operations are supported); otherwise, *InitialDP* or *EventReportBCSM,* depending on whether it is issued at a TDP or EDP.
T_Mid_Call	*TMidCall* (if DP-specific operations are supported); otherwise, *InitialDP* or *EventReportBCSM*, depending on whether it is issued at a TDP or EDP.
T_No_Answer	*TNoAnswer* (if DP-specific operations are supported); otherwise, *InitialDP* or *EventReportBCSM*, depending on whether it is issued at a TDP or EDP.
Termination Attempt Authorized	*TermAttemptAuthorized* (if DP-specific operations are supported); otherwise, *InitialDP*.

consistency with the rest of the ASEs in the table, for which reason they are included in quotes.

6.6.2.3 Application contexts. As with other constructs, the specification of the application contexts (ACs) relevant to the SSF/CCF-to-SCF and SRF-to-SCF interfaces differs from the specification of the ACs related to the SCF-to-SDF interface. Again, the former ACs are specified using the *MACRO* notation, while the latter ASEs are specified using the Information Object construct of the later version of ASN.1.

The *APPLICATION-CONTEXT* macro has two clauses that respectively identify the ASEs "consumed" by the initiator and responder.

TABLE 6.2   CS-1 INAP Application Service Elements (ASEs)

ASE	Associated operations	Direction
Activity-test-ASE	*ActivityTest*	SCF-SSF
Advanced-BCP-DP-ASE	*OMidCall, TMidCall*	SSF-SCF
Assist-connection-establishment-ASE	*EstablishTemporaryConnection*	SCF-SSF
Basic-BCP-DP-ASE	*OriginationAttemptAuthorized, CollectedInformation, AnalyzedInformation, RouteSelectFailure, OCalledPartyBusy, ONoAnswer, OAnswer, ODisconnect, TermAttemptAuthorized, TBusy, TNoAnswer, TAnswer, TDisconnect*	SSF-SCF
BCSM-event-handling-ASE	*RequestReportBCSMEvent*	SCF-SSF
	EventReportBCSM	SSF-SCF
Billing-ASE	*FurnishChargingInformation*	SCF-SSF
Call-handling-ASE	*HoldCallInNetwork, ReleaseCall*	SCF-SSF
Call-report-ASE	*CallInformationRequest*	SCF-SSF
	CallInformationReport	SSF-SCF
Cancel-ASE	*Cancel, CancelStatusReportRequest*	SCF-SSF
Charging-ASE	*ApplyCharging*	SCF-SSF
	ApplyChargingReport	SSF-SCF
Charging-event-handling-ASE	*RequestNotificationCharging*	SCF-SSF
	EventNotificationCharging	SSF-SCF
Connect-ASE	*Connect*	SCF-SSF
DP-specific-event-handling-ASE	*RequestReportBCSMEvent*	SCF-SSF
	OriginationAttemptAuthorized, CollectedInformation, AnalyzedInformation, RouteSelectFailure, OCalledPartyBusy, ONoAnswer, OAnswer, ODisconnect, TermAttemptAuthorized, TBusy, TNoAnser, TAnswer, TDisconnect, OMidCall, TMidCal	SSF-SCF
Generic-disconnect-resource-ASE	*DisconnectForwardConnection*	SCF-SSF
Non-assisted-connection-establishment-ASE	*ConnectToResource*	SCF-SSF
SCF-activation-ASE	*InitialDP*	SSF-SCF
SCF-call-initiation-ASE	*InitiateCallAttempt*	SCF-SSF
SCF-SRF-activation-of-assist-ASE	*AssistRequestInstructions*	SRF-SCF, SSF-SCF
Service-management-ASE	*ActivateServiceFiltering*	SCF-SSF
	ServiceFilteringResponse	SSF-SCF

TABLE 6.2 CS-1 INAP Application Service Elements (ASEs) (*Continued*)

ASE	Associated operations	Direction
Signalling-control-ASE	*SendChargingInformation*	SCF-SSF
Specialized-resource-control-ASE	*PlayAnnouncement, PromptAndCollectUserInformation*	SCF-SSF (SRF)
	SpecializedResourceReport	SRF-SCF
SSF-Call-Processing-ASE	*CollectInformation, AnalyzeInformation, SelectRoute, SelectFacility, Continue*	SCF-SSF
Status-Reporting-ASE	*RequestCurrentStatusReport, RequestEveryStatusChangeReport, RequestFirstStatusMatchReport*	SCF-SSF
	StatusReport	SSF-SCF
Timer-ASE	*ResetTimer*	SCF-SSF
Traffic-management-ASE	*CallGap*	SCF-SSF
"Connection-ASE" (defined as *dapConnectionPackage* information object of class *CONNECTION-PACKAGE)*	*Bind, Unbind*	SCF-SDF
"Search-ASE" (defined as *searchPackage* information object of class *OPERATION-PACKAGE)*	*Search*	SCF-SDF
"Modify-ASE" (defined as *modifyPackage* information object of class *OPERATION-PACKAGE)*	*AddEntry, RemoveEntry, ModifyEntry*	SCF-SDF

Technically, the ACs so defined (in Clause 2.1.5 of the Recommendation) are merely groups of ASEs.[274]

The *APPLICATION-CONTEXT* class used for the SCF-to-SDF interface (in Clause 2.2.2.2.2.1 of the Recommendation), in addition to the *CONTRACT* field, (see the preceding section), also contains the fields that specify dialogue mode, dialogue termination, and abstract syntaxes employed.

Table 6.3 lists the CS-1 INAP ACs together with the ASEs they contain. The name of the SCF-to-SDF AC is invented to keep the consistency with the rest of the AC, for which reason it is included in quotes.

---

[274]The ASEs have been grouped into ACs with the utmost care, so that they cover as many implementation options as envisioned by the companies involved in writing the standard.

TABLE 6.3   CS-1 INAP Application Contexts (ACs)

AC	Associated ASEs	Dialog Initiator
IN-CS1-SSF-to-SCF-Generic-AC	*SCF-activation-ASE, Assist-connection-estab-lishment-ASE, Generic-disconnect-resource-ASE, Non-assisted-connection-establishment-ASE, Connect-ASE, Call-handling-ASE, BCSM-event-handling- ASE, Charging-event-handling-ASE, SSF-call-processing-ASE, Timer-ASE, Billing-ASE, Charging-ASE, Traffic-management-ASE, Status-reporting-ASE, Call-report-ASE, Signalling-control-ASE, Specialized-resource-control-ASE, Cancel-ASE, Activity-test-ASE*	SSF
IN-CS1-SSF-to-SCF-DP-specific-AC	*IN-CS1-SSF-to-SCF-Generic-AC + Basic-BCP-DP-ASE, Advanced-BCP-DP-ASE, DP-specific-event-handling-ASE*	SSF
IN-CS1-assist-handoff-SSF-to-SCF-AC	*SCF-SRF-activation-of-assist-ASE, Generic-disconnect-resource-ASE, Non-assisted-connec-tion-establishment-ASE, Call-handling-ASE, Timer-ASE, Billing-ASE, Charging-ASE, Status-reporting-ASE, Specialized-resource-control-ASE, Cancel-ASE, Activity-test-ASE*	SSF
IN-CS1-SRF-to-SCF-AC	*SCF-SRF-activation-of-assist-ASE, Specialized-resource-control-ASE, Cancel-ASE*	SRF
IN-CS1-SCF-to-SSF-AC	*Assist-connection-establishment-ASE, Generic-disconnect-resource-ASE, Non-assisted-connec-tion-establishment-ASE, Connect-ASE, Call-handling-ASE, P-specific-event-handling-ASE, Charging-event-handling-ASE, SSF-call-pro-cessing-ASE, SCF-call-initiation-ASE, Timer-ASE, Billing-ASE, Charging-ASE, Status-reporting-ASE, Call-report-ASE, Signalling-control-ASE, Specialized-resource-control-ASE, Cancel-ASE, Activity-test-ASE*	
IN-CS1-SCF-to-SSF-traffic-management-AC	*Traffic-Management-ASE*	SCF
IN-CS1-SCF-to-SSF-service-management-AC	*Service-Management-ASE*	SCF
IN-CS1-SCF-to-SSF-status-reporting-AC	*Cancel-ASE, Status-reporting-ASE*	SCF
*"IN-CS1-SCF-to-SDF-AC"**	*"Connection-ASE" (dapConnectionPackage), "Search-ASE" (searchPackage), "Modify-ASE" (modifyPackage)†*	SCF

*Defined as the INDirectoryAccessAC information object of the *APPLICATION-CONTEXT* class.

†Note: according to the definition of the *APPLICATION-CONTEXT* syntax, the object of its class contains more than a collection of ASEs, which are listed above just to keep consistency with the rest of the CS-1 Application Context definitions.

6.6.3   Procedures

The procedures describe the sequencing of the protocol operations. In summary, Recommendation Q.1218 addresses the procedures in several places. First, Clause 1 of the Recommendation describes the rules that deal with TCAP AC negotiation and execution of operations.[275] In addition, Clause 2.2.2.2.1 describes similar rules regarding the use of the X.500 Directory protocol (on which the portion of INAP related to the SCF-to-SDF interface is based). Secondly, Clause 3 of the Recommendation provides the FSM-based model for the description of the INAP procedures and detailed specification of all involved FSMs. (Subsequently, the procedural description of all INAP operations specifies their pre- and postconditions in terms of the state model of Clause 3.) Finally, Annex A of the Recommendation contains the SDL description of the model of Clause 3.

Thus, Clause 3 is the primary place in the Recommendation to look for the INAP procedure's specification. The specification, however, appears to be overwhelming to begin with, and it is further complicated by the presence of mixed options. This happened because two conflicting (although both natural and necessary) considerations have been involved in shaping the Recommendation. One consideration is to constrain the service logic as little as possible and therefore make the prescribed sequencing of operations rather loose. The other is to ensure that the products manufactured by different vendors work in concert with each other. This consideration demands the clarity of protocol semantics, which is achieved through constraining the sequencing.

Once again, the authors would like to attract the reader's attention to Recommendation Q.1219, which is extraordinarily helpful in interpreting the procedures in relation to specific services and service features. In this book, we describe the framework of the procedural model and review just enough aspects of the specification for the SCF side to be able to continue with the interpretation examples of Service Assist/Hand-off and Queuing features that we have been following on throughout the book. The remainder of this section provides the necessary background on the protocol model followed by interpretation examples.

6.6.3.1   The modeling of IN entities.

The abstract model of an IN FE (which is applicable to all IN FEs) is presented in Fig. 6.4. This is *not* the model used in the Recommendation, however. The latter actually provides four different FE models (for the SSF, SCF, SRF, and SDF),

---

[275]The Recommendation specifically refers to these as SACF and MACF rules. These rules alone, however, do not form a complete set of SACF and MACF because they address only the interworking with TCAP rather than the semantics of INAP.

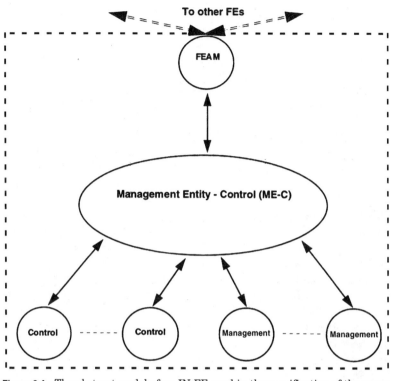

Figure 6.4   The abstract model of an IN FE used in the specification of the procedures.

but the present model is a generalization that covers the common aspects of those models.

The purpose of the model is to support the multitude of the events inherent to systems of communicating asynchronous processes. In this model, a *Functional Entity Access Manager* (FEAM) is responsible for the actual message exchange with other FEs. Each message received by the FEAM is passed to the *Management Entity–Control* (ME-C) object, which, in turn, passes it to its appropriate subordinate object. The subordinate objects include *Control* objects and *Management Entity* objects. (For example, in the SCF, the Control objects[276] receive the call-instruction-related operations, while the Management Entity objects[277] keep track of status requests.) The ME-C object creates and destroys its subordinate objects whenever such a need arises. (Again, in the case of SCF, a control object is created whenever an initial—or

---

[276]These objects are in the class *Service Control State Model* (SCSM).

[277]These objects are in the class *Service Control Management Entity* (SCME).

DP-specific—request for an instruction from the SSF/CCF arrives or when the SCF has to initiate a call by itself; the object is destroyed when the call needs no further IN support.) The ME-C object is also responsible for passing the messages from its subordinate objects to the FEAM.

This model is realized in all IN FEs. In other words, each IN FE is modeled as an object of the class defined by that abstract model. To this end,

1. Clause 3.1.1 of the Recommendation specifies the model for the SSF, where the ME-C is realized as the SSF management entity–control (SSME-Control), control objects are realized as SSF finite state machine (SSF-FSM) objects,[278] and management entity objects are realized as SSF management entity (SSME) objects.

2. Clause 3.1.2 of the Recommendation specifies the model for the SCF, where the ME-C is realized as the Service Control Management Entity–Control (SCME-Control), control objects are realized as the SCF Call State Model (SCSM) objects, and Management Entity objects are realized as the Service Control Management Entity Finite State Model (SSME-FSM) objects.

3. Clause 3.1.3 of the Recommendation describes the model for the SSF, where the ME-C is realized as the Specialized Resource Management Entity (SRME) and control objects are realized as the SRF Call State Model (SRSM) objects. (No management entity objects are presently needed in the SRF.)

4. Clause 3.1.4 of the Recommendation describes the model for the SDF. Here the model is very simple: only the SDF FSM object is defined (because no coordination of separate requests to the SDF needed to be modeled[279]).

It is important to understand that this model is relevant only to interfaces, and, as such, it may have nothing to do whatsoever with the internal structure of the FEs. In other words, when two objects communicate via a well-defined set of messages, each object needs to know how to react to messages that it receives from another object, but this knowledge does not have to represent the internal structure of its interlocutor.

Another important point to keep in mind is that although the Recommendation contains references to FSMs that receive or send

---

[278]The names of this object and its counterparts in other FEs were selected to parallel the BCSM.

[279]Alternatively, the model could be augmented by management entity objects, which would create an SDF FSM object on the reception of the *Bind* operation from the SCF, and destroy it when the *Unbind* is received.

messages, they, strictly speaking, do not. It is the objects that send and receive the messages, but they do it as prescribed by relevant FSMs.

There has always been some confusion about the purpose of the modeling with the FSMs in Recommendation Q.1218, especially regarding modeling within the SCF. For example, Clause 3.1.2.3 of the Recommendation mentions "the primitive interface between the SCF FSM and SLPs/maintenance functions," although it notes that this interface "is not a subject of standardization in CS-1." In reality, it is much better to think about the SCF FSM as a model of the service logic itself. A good metaphor for the relation of the FE models to the FEs themselves is the relation of a set of differential equations that model a piece of mechanical equipment to that equipment. It hardly makes sense, for example, to discuss the "interface" between the equations and the equipment. Nor does it makes sense to think of the differential equations as *the* equipment itself; they are there to describe certain aspects of the equipment's behavior. Similarly, the FE models describe the behavior of the FEs inasmuch as it is visible at the relevant interfaces. To this end, the SCF FSM describes generic service logic behavior as far as the interfaces to the SSF/CCF, SRF, and SDF are concerned.

In what follows, we review the SCSM.[280] We start with the Fig. 6.5 (a copy of Fig. 18/Q.1218), which depicts four high-level states of the SCSM. The adjective *high-level* indicates that each of these states can further expand into a separate, lower-level state machine. Before considering such expansions, however, we first familiarize ourselves with a few notational conventions and general working of the model.

The starting state of the SCSM is the *Idle* state. The active life of the SCSM object starts when it either receives a query from the SCF (event *(E2) Query_from_SSF*) or the service logic in the SCF starts a call on its own (event *(e1) SL_Invocation*). Note that the letter E is capitalized in the former case to indicate that the event is *external* (i.e., caused by a reception of a message); similarly, the names of the *internal* events always[281] start with the lowercase e inside the parentheses.

Through its life cycle, the SCSM is working on preparing the SSF instructions (the activity, which is carried in state 2). If a call party's data are needed to carry this work, the SCSM first attempts to obtain the specialized resource (in state 3), and, if successful, it supervises the user interaction (in state 4), on completion of which it returns to state 2. The SCSM is quite straightforward at this level, except for one transition that seems to be out of place, namely the transition that corre-

---

[280]The SCSM is by far the most complex of all models, which the reader should have no problem figuring out once he or she becomes comfortable with the SCSM.

[281]This convention holds for all models in the Recommendation.

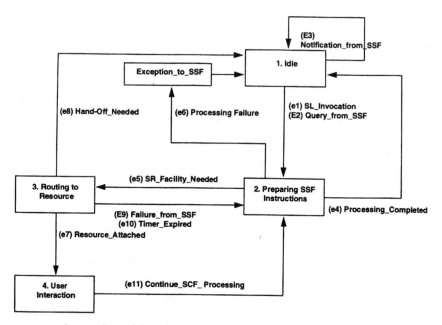

Figure 6.5 Service Control State Model (SCSM). (*After Fig. 18/Q.1218.*)

sponds to the event (*e8*) *Hand-Off_Needed*. Indeed, the transition moves the SCSM to the *Idle* state, which seems to be puzzling. We promise the patient reader to present the solution to the puzzle in Sec. 6.6.3.2.2. This is one example of the complexity inherent in the model.

Figures 6.6 (after Fig. 19/Q.1218) and 6.7 (after Fig. 21/Q.1218) furnish the expansions of states 2 and 3 of the SCSM, respectively.

These (and other) expansion figures should be interpreted as follows. All the transitions into the original state lead into the starting state of the expansion FSM; each transition out of the expansion FSM (contained within a dotted rectangle) maps into an event of the original FSM (depicted within the parentheses). Any state of an expansion FSM can, of course, be expanded again, which is what happens with state 2.2 (Fig. 6.7). This state is further expanded into a Queuing FSM, as depicted in Fig. 6.8 (replicating Fig. 20/Q.1218).

Overall, the FSMs of Recommendation Q.1218 have one problem: they are nondeterministic. While it has been ascertained that every sequence of messages anticipated by the contributors to the standard can be derived from the FSMs, some unanticipated sequences can be so derived, too. Selecting proper options does pin the selection of sequences down, but the options selected by different developers may be incompatible with each other. To help the reader understand the key issues involved in selecting options, the next section contains several practical interpretation examples relevant to existing implementations.

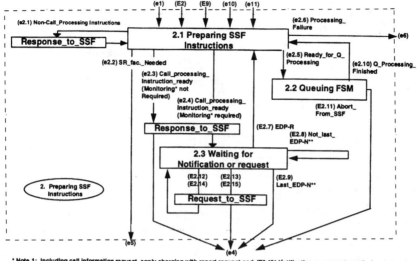

(e1) (E2) (E9) (e10) (e11)

(e2.1) Non-Call_Processing Instructions

Response_to_SSF

2.1 Preparing SSF Instructions

(e2.6) Processing_ Failure

(e6)

(e2.2) SR_fac._Needed

(e2.5) Ready_for_Q_ Processing

(e2.10) Q_Processing_ Finished

2.2 Queuing FSM

(e2.3) Call_processing_ instruction_ready (Monitoring* not Required)

(e2.4) Call_processing_ instruction_ready (Monitoring* required)

Response_to_SSF

(E2.11) Abort_ From_SSF

(E2.7) EDP-R

(E2.8) Not_last_ EDP-N**

2. Preparing SSF Instructions

2.3 Waiting for Notification or request

(E2.12) (E2.13)
(E2.14) (E2.15)

(E2.9) Last_EDP-N**

Request_to_SSF

(e5)

(e4)

* Note 1: Including call information request, apply charging with report request and Request notification charging event

** Note 2: Including call information report, apply charging report and event Notification charging

(E2.12) Notification_or_request_continuing_instruction
(E2.13) Monitoring_cancel_instruction
(E2.14) Release_call_instruction (call information report Or apply charging report has been requested)
(E2.15) Release_call_instruction (neither call information Report not apply charging report has been requested)

Figure 6.6 Expansion of state 2. (*After Fig. 19/Q.1218.*)

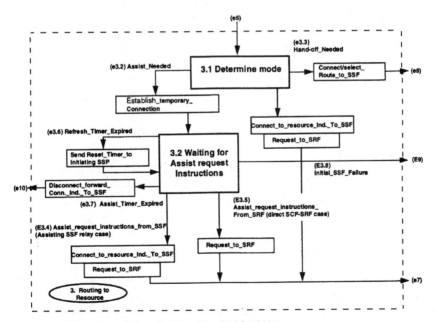

(e5)

(e3.3) Hand-off_Needed

(e3.2) Assist_Needed

3.1 Determine mode

Connect/select_ Route_to_SSF

(e8)

Establish_temporary_ Connection

Connect_to_resource_Ind._To_SSF

Request_to_SRF

(e3.6) Refresh_Timer_Expired

3.2 Waiting for Assist request Instructions

Send Reset_Timer to Initiating SSP

(E3.8) Initial_SSF_Failure

(E9)

(e10)

Disconnect_forward_ Conn._Ind._To_SSF

(e3.7) Assist_Timer_Expired

(E3.5) Assist_request_instructions_ From_SRF (direct SCF-SRF case)

(E3.4) Assist_request_instructions_from_SSF (Assisting SSF relay case)

Connect_to_resource_Ind._To_SSF

Request_to_SRF

Request_to_SRF

3. Routing to Resource

(e7)

Figure 6.7 Expansion of state 3. (*After Fig. 21/Q.1218.*)

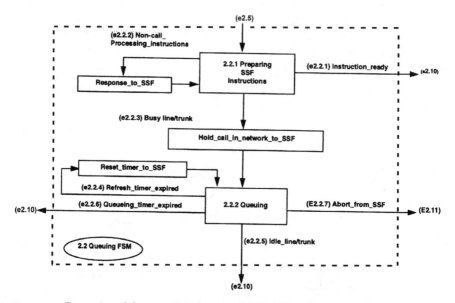

Figure 6.8 Expansion of the state 2.2 (Queuing FSM). (*After Fig. 20 / Q.1218.*)

6.6.3.2 Interpretation examples.

In the two examples chosen for this section we review the procedures for the two capabilities that are most complex as far as the procedures are concerned. In doing so, we offer an *interpretation* of a standard; in other words, we describe a set of options that transform the nondeterministic FSMs of Recommendation Q.1218 into deterministic ones (but in a way that is consistent with the standard). Both interpretations are based on the realization of these capabilities in the AT&T network in support of advanced 800 service.

Both capabilities are familiar to the reader. The first one is Call Queuing, whose Stage 1 description was first discussed at length in Chap. 4; the second is the Service-Assist capability discussed in Chap. 5.

6.6.3.2.1 Call Queuing in the network.

We recall that the Call Queuing feature allows the calling party to be kept on hold, possibly entertained by announcements, while the persons responsible for answering the telephones (often called the *agents*) are busy with other calls. When an agent is free, the calling party that called the earliest is connected to the agent. Thus, the calls are queued according to the "first-come, first-served" discipline.

A general configuration is depicted in Fig. 6.9, where a particular call started via an SSP (which issued a request for instruction to the SCP) may be answered by the agents in different locations served by different SSPs and possibly PBXs. As the figure demonstrates, the calls span several networks.The SSPs are connected through the SS No. 7 net-

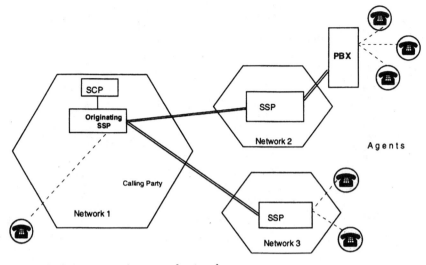

Figure 6.9 Service spanning several networks.

work, but there is only one SCP through which the services are offered. The SCP is connected to the SSP through which the calls enters the network (network 1) where the services are provided.[282] (Neither the trunks connecting the SSPs with each other, nor the intermediate SSPs, nor the IPs are shown.)

In order to complete calls, the service logic must be aware of the status of the agents' lines. In other words, those have to be monitored. To this end, Clause 3.1.2.5.2.2 of Recommendation Q.1218 considers four possibilities for resource monitoring, which are started by the following operations:

- *RequestFirstStatusMatchReport*
- *RequestCurrentStatusChangeReport*
- *RequestEveryStatusChangeReport*
- *RequestEventReportBCSM*[283]

This is the first call for selecting an option for a specific implementation. We observe that the first three cases work *only* when the agents' lines are attached to the *same* SSP that issued the initial request.

---

[282]Even though Fig. 6.2 depicts (for simplicity) the configuration where the SSP is the first to receive the call from the call party, this need *not* be the case in general. In fact, it has not been the case in the AT&T Long Distance Network implementation, where the SSPs (No. 4 ESS™ switches) receive calls through the SSPs in local networks.

[283]Specifically, the *O_Disconnect* and *O_Abandon* events.

Indeed, the SCP may issue the monitoring operations only to those SSPs with which it has established relations, but it would not be efficient to establish relations with all the terminating switches. In fact, doing so may even conflict the single-endedness principle of CS-1.

The fourth option, however, results in monitoring only for the duration of a single call and—even more importantly—only at the originating SSP. As the parallel discussion of queuing in Chap. 4 pointed out, the only price to pay for the simplicity is the requirement that (1) *only service-related calls be directed to the agents* and (2) *agents may not originate calls.*

Again, selecting the fourth option allows the network to avoid continuous monitoring of calls: a switch sends to an SCP a notification reporting a change (from busy to idle) in the status of a (likely remote) circuit only when it has been asked in advance by an SCP to do so. This can be implemented using the mechanism used in operating systems and known as *context switching.*[284]

We start by shedding some light on the *Resource Control Object* (RCO), an otherwise obscure entity of Clause 3.1.2.4.5 of Recommendation Q.1218. The RCO is defined there as an object within the SCF—the *SCME Control,* to be precise—that keeps track of the available lines and trunks. (Initially, these are assumed to be free.) The RCO data reside within the SDF. The Recommendation lists the three *methods* through which these data are accessed: *Get_Resource, Free_Resource,* and *Cancel.* Clause 3.1.2.4.5 defines these methods as follows:

1. *Get_Resource.* This method is used to obtain the address of an idle line on behalf of an SCSM. If the resource is busy, the SCSM is queued for it.

2. *Free_Resource.* This method is used when a disconnect notification from the SSF is received. The method either advances the queue (if it is not empty) or marks the resource free (otherwise).

3. *Cancel.* This method is used when either the queuing timer has expired or the call has been abandoned.

Figure 6.10*a, b,* and *c* contains the respective SDL diagrams for these methods.[285]

We will return to the RCO whenever we need to invoke its methods, and we will demonstrate how it works with the SCSM. Presently, we

---

[284]This analogy has been explored in Faynberg et al. (1992) and applied to the CS-1-refined model in Dacloush et al. (1995).

[285]These diagrams are *not* part of the Recommendation; they represent our interpretation of the standard.

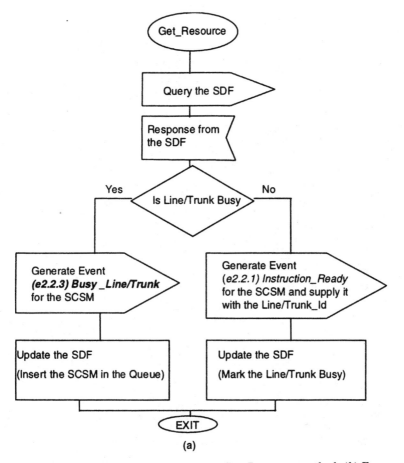

Figure 6.10 Resource control object. (*a*) Get_Resource method; (*b*) Free_
Resource method; (*c*) Cancel method.

consider state 2 of the SCSM (Fig. 6.5), in which the SCF is preparing
the routing instruction for the SCF. This state is expanded in Fig. 6.6,
where the *SCSM* starts in state 2.1. The only event relevant to the discussion is that of (*e2.5*) *Ready_for_Q_Processing*.[286] This event results
in the transition to state 2.2, which is, in turn, expanded in Fig. 6.8.

Now, the SCSM is in state 2.2.1, where the SCF is preparing the
instruction for the SSF to complete the call. To obtain the routing
address, the SCF invokes the *Get_Resource* method of the RCO.
Consulting Fig. 6.10(*a*), we observe that the method either finds a free

---

[286]This event corresponds to the decision made within the service logic to invoke a
branch that supports the call queuing capability.

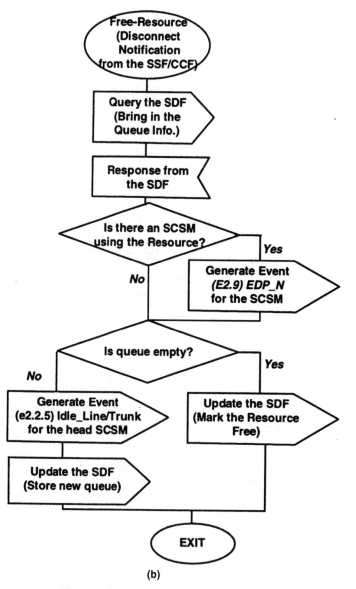

(b)

Figure 6.10 (*Continued*)

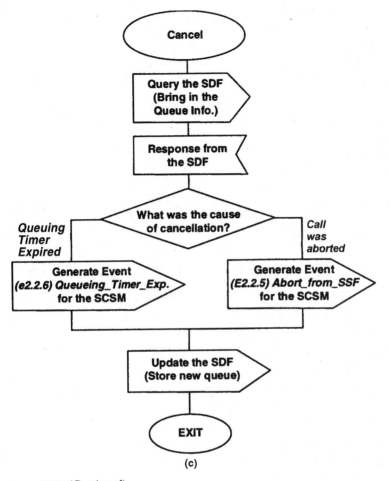

Figure 6.10 *(Continued)*

resource, generating event *(e2.2.1) Instruction_Ready* for the *SCSM,* or places the *SCSM* in the queue, generating event *(e2.2.3) Busy_Line/Trunk* for the *SCSM.*

If the resource is unavailable, the latter event causes the SCSM to send the *HoldCallInNetwork*[287] operation to the SSF/CCF, followed by a transition to state 2.2.2, *Queuing.*[288] Two timers are employed in this state: one registers the overall time the call spends in the queue, while

---

[287]In the ETSI Core INAP, this operation is omitted. Instead, the ResetTimer operation is sent.

[288]While in state 2.2.2, the SCSM may actually perform the work done in its states 3 and 4 to support the user interaction. It is a straightforward exercise to augment the queuing FSM with the appropriate states and transitions.

the other clocks the time interval since the last operation to the SSF/CCF was issued.[289] When the former timer (whose value is set by the service provider to control the associated expenses) expires, the call can no longer be kept in queue. In this case the Cancel method of the RCO is invoked. When the latter timer expires, the SCSM issues the *ResetTimer* operation to the SSF/SCF and restarts the timer. Going back to the *Cancel* method, it is also invoked when the call is aborted.[290]

If the resource is available, the queuing FSM is exited via the event (*e 2.10*), which causes the transition to the state 2.1 (Fig. 6.6). This is where the situation is becoming complex because of the nondeterminism of the *SCSM*. For the queuing—without continuous monitoring—to work, the event (*e2.4*) *Call_Processing_Instruction_Ready* (*Monitoring Required*) must take place.[291] This event causes the *RequestEventReportBCSM* operation together with the appropriate SSF instruction, which may contain routing and charging information, to be sent to the SSF as part of the *Response_to_SSF* action, and a transition to state 2.3, *Waiting for Notification or Request.* The BCSM event to wait for is *O_Disconnect,* and, for the purposes of the implementation in question, there should be no other outstanding EDP_N or EDP_R. On the reception of the disconnect notification, the SCME invokes the *Free_Resource* method of the RCO. It is the *Free_Resource* method that performs the context switching. Figure 6.10(*b*) demonstrates how it is done:

1. The SCSM that requested notification exits via the path indicated by the event (E2.9) Last_EDP-N, which maps into the event (*e4*) *Processing_Completed.*

2. The call that had been waiting at the head of the queue—if the queue is not empty—is handed a resource and thus becomes active [i.e., the corresponding SCSM in state 2.2.2, *Queuing* receives event (*e2.2.5*) *Idle_Line/Trunk*]. If the queue is empty, the resource is marked free.

6.6.3.2.2.   Service Assist and Service Hand-off.   We should mention that Annex A of ITU-T Recommendation Q.1219 contains a detailed inter-

---

[289]The SSF/CCF keeps its own timer to register that interval. If this timer expires, the SSF/CCF considers the relationship with the SCF nonexistent.

[290]Unless the appropriate DPs (e.g., O_Abandon) are set as TDP-N, an implementation may consider setting an EDP-N at the same time the HoldCallInNetwork operation is issued.

[291]The word *must* points to the appropriate implementation option to be chosen.

pretation of these capabilities at both the DFP and Physical Plane levels. The present section addresses these capabilities only from the point of view of the SCF procedures. In other words, this example—as well as the previous one—demonstrates how the capabilities may be programmed in the SCP, while Recommendation Q.1219 deals with the networkwide interpretation. The latter includes interactions with the SRF and SSF objects, without which the picture cannot be complete. (We note, however, that both the SSF and SRF models are simple compared to the SCF one.)

Figure 6.11, which the reader should find similar to Fig. 5.12, depicts the exchange of operations between the involved PEs in support of the Service Assist capability. The related activities are carried in states 3 and 4 of the *SCSM* (Fig. 6.5). The starting state is state 3, *Routing to Resource,* expanded in Fig. 6.7. The starting state of the expansion is state 3.1, *Determine Mode,* in which the service logic determines the SRF resources needed to support the service. The situation where such

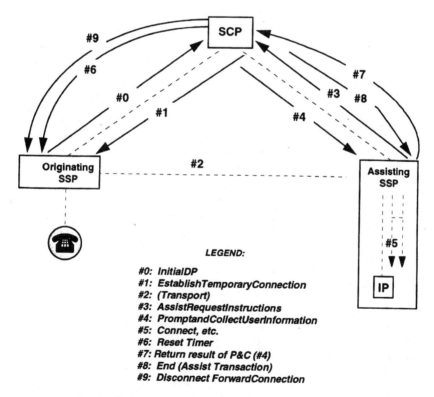

LEGEND:

#0: InitialDP
#1: EstablishTemporaryConnection
#2: (Transport)
#3: AssistRequestInstructions
#4: PromptandCollectUserInformation
#5: Connect, etc.
#6: Reset Timer
#7: Return result of P&C (#4)
#8: End (Assist Transaction)
#9: Disconnect ForwardConnection

Figure 6.11 Service Assist sequence.

resources are not present in the originating SSF/CCF[292] is reflected in the events (e3.2) *Assist_Needed* and (e3.3) *Hand-off_Needed.*

Let us start with the event (e3.3), whose handling is especially tricky. In this case, the SCF instructs the originating SSF/CCF to establish a connection to the assisting SSF/CCF that should last for the duration of the call. (This is done by issuing either the *Connect* or *SelectRoute* operation, which explains the name of the corresponding action, *Connect/Select_Route_to-SSF,* used in the *SCSM* expansion of Fig. 6.7.) At this point, the SCF terminates the relationship with the originating SSF/CCF, and the *SCSM* returns to state 1, *Idle.* When the request for the instruction arrives from the handoff SSF/CCF, the *SCME-Control* will create another instance of the *SCSM.* [Finally, we followed up on our promise earlier in this chapter to provide the solution to the puzzle with the event (e8) of the *SCSM.*]

Going back to the Assist case, we observe that the event (e3.2) causes the *EstablishTemporaryConnection* operation, which carries the routing address of the assisting SSP[293] to be issued to the originating SSF/CCF. On this event, a transition to state 3.2, *Waiting for Assist Request Instructions,* takes place.

In state 3.2, two timing activities are started. The activities are similar to those performed in call queuing, and they are carried for exactly the same reason. Two timers are employed: one registers the time the SCF spends waiting for the remote connection establishment, while the other clocks the interval since the last operation to the SSF/CCF was sent. When the former timer expires, the SCF terminates the assist procedure (by issuing the *DisconnectForwardConnection* operation to the SSF/CCF). When the latter timer expires, the *SCSM* issues the *ResetTimer* operation to the SSF/SCF and restarts the timer; this procedure is repeated for the duration of the user interaction.

When the *AssistRequestInstructions* operation is received by the SCF, the SCSM moves to state 4, *User Interaction.*[294] The termination procedure for Service Assist (which includes sending the *DisconnectForwardConnection* to the originating SSF/CCF and an

---

[292]In other words, the switch neither contains nor has an access to an IP with the needed resources.

[293]The standard also allows an IP to act as an assisting SSP.

[294]Again, this operation may be sent by either the assisting SSF/CCF, as reflected in the event (E3.4) *Assist_Request_Instruction_from_SSF,* or the SRF, as reflected in the event (E3.5) *Assist_Request_Instruction.* The only difference in handling these events is that the former results in issuing the *ConnecttoResource* operation ahead of the request to play announcements (i.e., *PlayAnnouncement* or *PromptAndCollectUserInformation* operation).

unspecified procedure for terminating the association with the assisting SSP[295]) is carried in this state.

We intentionally omitted a discussion of state 4; the reader should have no problem filling in the details by reading the Recommendation.

6.7 Draft Recommendation Q.1225, Intelligent Network—Physical Plane for CS-2

Figure 6.12 (which is a copy of the ever-changing Fig. 1/Q.1225) depicts the current Physical Plane architecture.

---

[295]The example in Recommendation Q.1219 uses the TCAP *End* operation to close the transaction with the assisting SSP, which is what is done in the AT&T and some other implementations; however, this method has not been prescribed by the standard.

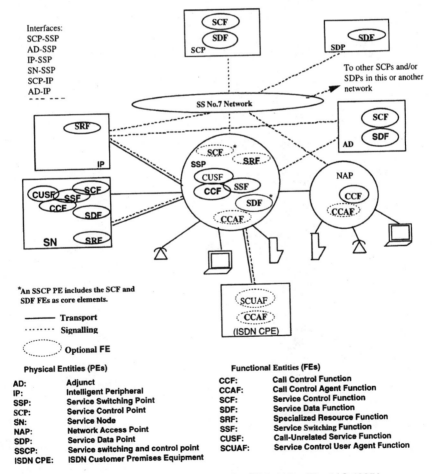

Figure 6.12   Draft Physical Plane architecture for CS-2. (*After Fig. 1/Q.1225.*)

First, note that a new PE, ISDN *Customer Premises Equipment* (ISDN CPE) has been introduced as follows: "ITU-T Recommendation I.112 defines the ISDN CPE as 'a node that provides the functions necessary for the operation of the access protocols by the user.' Functionally, the ISDN CPE can contain the SCUAF (for the bearer-unrelated interactions) and the CCAF." (The latter FE is presently optional.) The ISDN CPE has one interface—to the SSP. It is the only PE that contains the SCUAF.

The wireless FEs have not been mapped into PEs for CS-2. The only new FE left after the exclusion of the wireless FEs is the Call-Unrelated Service Function (CUSF). The CUSF has been assigned as a mandatory FE to both the SSP and SN.[296] That means that the CS-2-compliant switching equipment must support call-unrelated signaling.

---

[296]There has also been a proposal, which is likely to be accepted in the near future, to introduce the *Call-Unrelated Service Point* (*CUSP*), which contains the *CUSF,* into the Physical Plane architecture.

Current IN Standardization Work

7.1 Overview

Now that stable IN standards for the fixed (wireline) narrowband networks exist, the telecommunications industry wishes to extend them—using the principles of IN—toward supporting B-ISDN and wireless networks. Yet another dimension for IN growth is service creation.

With all that, the integration of computing and telecommunications is advancing so rapidly that new paradigms, which envision highly modular and portable software that manages standard switching fabrics, seem less and less futuristic. The new technology, once proven in the research laboratories and scrutinized by various industry forums, quickly finds its way into vendors' product plans. Andrew S. Tanenbaum cites (in Tanenbaum, 1989) an interesting and illustrative theory of standards development, which he calls the "Apocalypse of the Two Elephants." According to the theory, the capital invested into a new technology is first spent on the research and then on the development of the products. In each of these two periods the amount invested first grows very fast, then remains constant for a while, and then declines. When depicted graphically, each period is represented by a curve suggestive of an elephant. First the research elephant appears, then the development elephant. According to the theory, the standards should be written in the period separating the elephants (lest the standards people get crushed). It suffices to say that in the area of computing and telecommunications integration—which is what IN is all about—these two elephants are close enough to each other to present a challenge to the standards people. But even more challenge comes from the whole herd of the elephants that emerged from other standards bodies and forums and now are fast approaching IN (see Fig. 7.1).

The extent of the challenge will be clear when we highlight the current developments in the standards. First, we address the two aspects

Figure 7.1 The elephants approaching.

of the current work in Study Group 11 of ITU-T: the plans and progress of Capability Set 3 (CS-3) and the work on what used to be called the "Long-Term IN Architecture."[297] Then we review the work on the IN standards in regional standards bodies.

7.2 CS-3—The Starting Point

7.2.1 Summary

The CS-3 work quietly started in the summer of 1995. It has been over-shadowed by the work on CS-2, and will not progress at full speed until CS-2 is finished, which is expected to happen in the first quarter of 1997. The task of CS-3 is to develop the protocol that will support both Broadband ISDN (B-ISDN) and Wireless Personal Communications Services. CS-3 is also expected to standardize service creation. And—if that were not enough—CS-3 may pick up the discipline of Telecommunications Management Network (TMN) produced by Study Group 4 of ITU-T in order to specify the IN managed objects. This involves the integration of Network Traffic Management and Service Management (i.e., the areas traditionally covered by Operations Systems).

---

[297]This project is no longer called "Long-Term IN Architecture," but the new name has not been agreed on at the time this book is published. Most likely the name of the project will contain the words "information" and "networks," which is not surprising since the major contributor to the project will be the Telecommunications Information Networking Architecture Consortium (TINA-C).

Among the key services supported in CS-3 are Multimedia and Personal Communications Services (PCS), which cover both personal mobility [such as Universal Personal Telecommunications (UPT), already supported in CS-2] and terminal mobility.

Even though CS-3 has had very few meetings so far, by April 1996 it had gathered over 100 pages of material. At this point, predicting what *might* be part of the CS-3 Recommendations is pointless. The best the authors can do is highlight several new ideas and developments in computing that were brought into IN groups as the results of research.

7.2.2 Integration of IN and B-ISDN

As INAP is a data communications protocol, it is not inherently concerned with the bandwidth of the bearer connection; nor is it directly affected by the physical characteristics of any medium involved in a call. In theory, in order to support B-ISDN, it just has to adopt a number of operations and parameters, and its procedures have to evolve to coexist with the rest of the network activities.

There are two approaches to carrying the INAP modifications. One is a top-down approach, according to which all the objects relevant to the call first have to be defined (these objects will then compose a new call model). The new model itself can unify the TMN model with that of B-ISDN. Once the model is settled, it is expected to drive the development of the protocol naturally: whenever an object needs to send a message to another object, the data carried in that message are to be selected according to the intent of sending the message. So far, the material proposing various call models has been contributed to CS-3, but this material still remains too abstract to see a direct path that will link it with the development of the protocol.

The other is a bottom-up approach, based on juxtaposing the narrowband ISDN access architecture and standards with those of B-ISDN. The proponents of this approach note that as long as INAP is essentially a protocol between the switch and service control, its messages should be populated with the data defined in the basic call. The basic call data in a narrowband ISDN switch are affected by the messages that arrive to it from either the *Customer Premises Equipment* (CPE) via the access protocol (e.g., Q.931), or from another ISDN switch via the ISUP protocol. Conversely, if the SCP is to start a new call or influence the existing one, the messages that it is to send to the switch must contain the information that can be carried either to the CPE or to another switch. Then all that has to be done in order to define the broadband INAP is examine the B-ISDN access protocols and B-ISUP and add the necessary parameters to INAP.

As simple and straightforward as this approach appears to be, it cannot be carried out right away without encountering some difficulties.

First, even in the case of the CS-1 INAP, the states of the access call model are not completely aligned with that of BCSM, and not all INAP-carried data are compatible with those of the access protocols and ISUP.[298] Secondly, B-ISDN is also evolving, which means that some aspects of its protocols keep changing.

It is natural to expect then that both the top-down and bottom-up approaches will be carried simultaneously up to the point when their results will collide with each other. At that point, as is usually done in standards, the "harmonization" process will be carried, in which the results of both processes will be revised iteratively until the common ground is found.

7.2.3 Distributed computing (Middleware)

Traditionally, software applications have been written on top of *operating systems* whose major functions were—and still are—to support a specific *Application Programming Interface* (API) necessary for the development of applications as well as provide the necessary run-time support to the application processes. The latter includes scheduling, synchronization, uniform file access, interprocess communications, memory allocation, etc. In fact, the work that the operating systems are performing is associated with this second function, even though programmers, in general, are concerned only with API—the rest is implicit. In the previous software generation, the operating systems provided APIs for interprocess communications, but those included only communications among the processes on one machine. The software for intermachine communications was written later; it was usually added on top of operating systems. Because this communications software was written without any specific standard in mind (the OSI Application Layer had gained substantial material only in the late 1980s), the integration of different platforms was hard to achieve and much software had to be rewritten in order to do so.

With the advance of distributed communications and evolution toward *location transparency* (i.e., the application requesting a resource, say a file, being unaware of the specific machine on which this resource may reside), network operating systems have been introduced. In a network operating system, the users are aware of the existence of multiple computers; they can log in to remote machines and copy files from one machine to another. The API provided by such systems also requires explicit naming of the machines and resources. A distributed operating system, in contrast, is one that appears to application programmers as a traditional uniprocessor system, even though

---

[298]There is a protocol interworking group within the ITU-T Study Group 11 that has been addressing these issues.

it actually runs on multiple machines. In a true distributed system, neither application programmers nor the users of applications should be aware of where their programs are being run or where their files are located; that should be all handled automatically and efficiently by the operating system. The major characteristic of such systems is the presence of a single global interprocess communication mechanism. In terms of API, this amounts to a single set of system calls available on all machines.

Although a considerable body of knowledge about designing distributed operating systems is now available, the industry's economic considerations still make major computer vendors produce proprietary single-machine operating systems as part of their equipment. Consequently, in a multivendor environment, a distributed operating system is not a reality. A close alternative, however, is to develop an adaptation layer of software *on top* of *all* operating systems so that this layer provides a uniform API similar to what a distributed operating system would provide. This layer is what is called *Middleware.* The programming interface that allows Middleware to adapt to a particular operating system is called *Systems Programmer's Interface* (SPI). Naturally, different operating systems have different SPIs. On the other hand, even if the Middleware adapts to only one operating system, it still provides a nontrivial and important capability, by hiding the distribution from its users. Of course, an SPI is the same interface that is available to application programmers (as an API); perhaps, the name SPI was chosen to indicate that Middleware is a complex and versatile system. Figure 7.2 depicts the Middleware architecture.

The present definition of middleware in the CS-3 Baseline Document is as follows: "Middleware is a layer of functionality residing above

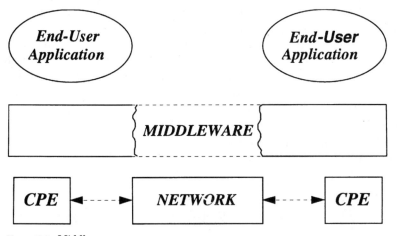

Figure 7.2 Middleware.

operating system and low-level networking software used by distributed applications to enable inter-operation."

The capabilities available to users of middleware are called *middleware services*; the ones defined in the CS-3 Baseline Document are as follows:

Context Service. This is a capability to create, change, and destroy contexts. A *context* is a blueprint of a particular type of a call, which describes the involved parties, their connections, etc. Thus a context is created before a call starts and it exists after the call finishes. The communicating parties are joined as members of the context. Each member of a context is represented by a context agent, which negotiates the creation of context with the context server.

Name Service. This is a capability to assign names to the resources and other entities that compose contexts (users and servers alike), which—at run time—are mapped into their network addresses. Even if the address of a person or a resource associated with a context changes, the Name Service keeps this change transparent to the members of the context.

Trader Service. This capability allows the service programs to advertise their services (by listing certain properties of these services), and it allows other programs to solicit such services (by providing their properties). At run time, the Trader Service matches the requests with the servers.

Virtual Transport Service. This capability provides a single API for communications among context servers and context agents and among context servers and other context servers.

A reader familiar with the *Open Distributed Processing* (ODP) Reference Model (see ISO, 1993) will find these services familiar. The Middleware material applies ODP to specific IN tasks. The Baseline CS-3 Document contains an example of a complex multimedia call setup, which involves (1) a user, John, whose telephone is connected to a PBX guided through a call control interface, (2) another user, Jane, whose telephone is connected to a multimedia workstation which sets up voice calls over a Local Area Network (LAN), and (3) several pieces of the audio and video equipment interconnected by routers and a video conference bridge. The example demonstrates how the application interacts with the Middleware services in order to set up a multimedia call:

1. John's application invokes the Context Service. As the result, John's context agent requests that the Context Server create a context for the call.

2. John's application invokes the name service in order to locate Jane's context agent.

3. John's application, again using the context service, requests that Jane's Context Agent (armed with the audio and video capabilities) join the context.

4. John's and Jane's audio agents exchange negotiation messages to establish a voice connection between their telephones.[299]

5. John's audio agent initiates (via the telephony server) the call to Jane.

6. John's and Jane's video agents exchange negotiation messages to establish a bearer channel for the video stream.

7. John's and Jane's video agents request that their respective video device establish a connection to the video server on the video conference bridge.

As the reader may observe, programming the above scenario—providing that the middleware is in place—would not demand much more effort than the prose description of the example.

7.2.4 Service Creation (service logic portability)

The discussions on Service Creation and Service Management are held in a separate group, which publishes the results of its studies in the *Service Management Baseline Document.* As the scope of a particular Capability Set crystallizes, the material from that document migrates to relevant Recommendations. The material discussed in this section is presently destined for CS-3.

First we should clarify what service creation standardization is all about; for one thing, its subject is very different from protocol definition. Standardization of service creation really means standardization of particular languages in which service programs are written as well as their respective programming environments. Many vendors of service creation environments feel that their products give them strong competitive advantages, and so they prefer to keep both—the languages and the programming environments—proprietary.

The service providers would prefer a standard in service creation, but their major problem is not an absence of such a standard, but the

---

[299]Negotiation messages flow from John's audio agent to his context agent, to the context server, to Jane's context agent, and then to her audio agent. This results in the agreement to establish a voice path by way of the telephony server telephone device controller on John's endpoint and the LAN phone device controller on Jane's endpoint.

cost of maintaining and reusing service logic programs in a multiven-dor environment. A service provider can always develop service logic programs in one specific environment as long as the equipment in the provider's network can execute these programs. It is not quite easy to achieve that, if different pieces of the equipment are bought from different vendors. Furthermore, when a service creation environment, which is better and more versatile than what a provider has had, appears on the market, the provider may wish to purchase it, very often together with a new SCP or SN. But what does one do with the services that had been developed in the old service creation environment? Reprogramming them in a new environment is likely to turn out to be prohibitively expensive. What about keeping the old service creation environment just to maintain the services that had been developed? It is likely that the new SCP will not run the old Service Logic Programs—the service creation environments are usually tightly cou-pled with the service execution platforms.

In 1992, AT&T assigned Bell Labs researchers to study this problem. The assumption was (and it is still correct today) that neither a service creation environment nor the instruction set of a service execution environment are likely to be standardized in the near future. The result of the research,[300] complete with the convincing feasibility pro-totyping effort, proposed a service creation architecture that supports multiple service creation paradigms and multiple execution environ-ments. This architecture, however, can be realized in a multivendor environment *only* if the vendors agree to standardize a certain aspect of it. The introduction of this material into standards has so far gener-ated enough interest and received enough support to warrant its inclu-sion into the Service Management Baseline Document.

Before addressing the proposed architecture, we discuss the service creation process (see Fig. 7.3).

Service logic programs are written in specific—often proprietary—languages, which are then translated by compilers into the target code executed by a particular processor.[301] In terms of the IN PEs, such a processor can be either an SCP, IP, SSP, SN, or all of them, if the lan-guage is versatile enough to specify distributed processing so that its compiler can generate the target code for all processors involved. Consequently, in order to port the service logic produced in m service creation environments to n different processors (see the left part of Fig. 7.3) with the brute-force method, $m \times n$ compilers are needed. The development of a compiler, however, is a major software undertaking

---

[300]See Slutsman et al. (1994).

[301]The target code can take form of the machine-language instructions, in which case it is run directly by the processor, or it may be *interpreted* at the run time by an *inter-preter* process that runs on that processor.

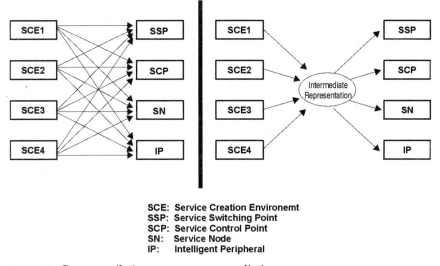

SCE: Service Creation Environemt
SSP: Service Switching Point
SCP: Service Control Point
SN: Service Node
IP: Intelligent Peripheral

Figure 7.3 Cross-compilation versus common mediation.

that requires significant investment of both time and money. If, however, the compilers produced their output in some uniform intermediate representation, which then can be translated into the target code (as depicted in the right side of Fig. 7.3), then only $m + n$ translators are needed. The latter approach (called *common mediation*) promises a significant cost improvement (linear versus quadratic) over the former one, if it can be done, of course.

The research proved that it can indeed be done, as follows. All compilers first go through a phase called *parsing,* in which the syntactic structure of the program to be translated is determined. At this stage, the compilers produce *parsing trees* of the programs. In the next phase, the compilers generate the target code. Thus, every existing compiler can be modified to output a program's parsing tree in an appropriate format, and (relatively inexpensive) *parsers* can be written for this purpose, too. The code generators for different processors can be written independently from parsers. Those would pick up parsers' output and produce the target output. For all of that to work, the format of the parsing tree must be standardized. The *Application-Oriented Parsing Language* (AOPL) is proposed to serve this purpose (as summarized in Fig. 7.4).

Presently, the AOPL grammar and methodology are discussed in standards. It should be noted, however, that using AOPL would make sense economically only if at least three service creation languages were simultaneously used over more than two service creation platforms. It is up to the industry to decide whether AOPL is needed—and, therefore, needs be standardized—in the near future.

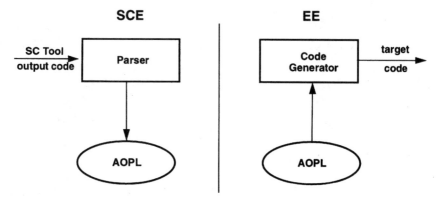

Figure 7.4   Two steps in porting the service via AOPL.

7.3   ITU-T "Long-Term Architecture"

Again (see footnote 297), this project has been initiated in ITU-T, although its name has not been agreed on yet. The foundation of the project is the output of the Telecommunications Information Networking Architecture Consortium (TINA-C), which was brought into ITU-T by the member companies. Since almost all companies that are involved in TINA are members of ITU-T, and, conversely, most companies that are active in ITU-T are members of TINA-C, it is likely that the output of TINA-C will become the future IN standard.

The rest of this section introduces TINA-C, describes its relation to IN, and then proceeds to a high-level overview of the TINA architecture.

7.3.1   TINA-C

The Telecommunications Information Networking Architecture Consortium (TINA-C) was formed in 1993. It is the principal forum within the telecommunications industry for precompetitive, prenormative research and specification generation in the area of distributed computing as applied to Telecom industry concerns. TINA-C is engaged in specifying and validating an architecture to support future telecommunications and information services based on advanced distributed computing concepts. The members of TINA-C include most of the world's large global carriers, as well as Telecom and computer vendors. The main work of the Consortium is carried out by a Core Team of about 40 researchers contributed from the member companies. This Core Team is colocated at a Bellcore facility in New Jersey.

While TINA-C's own work is by definition "prenormative," the Consortium seeks to influence other consortia and traditional standards bodies, including the Object Management Group, ATM Forum, and, closest to the specific subject of this book, the longer-term aspects of the IN work in the ITU.

7.3.2 Relationship between TINA-C and IN

Actually, there are many interesting relationships between TINA-C and IN. At the level of objectives, TINA-C and IN appear to be quite similar. Both promise to help network operators and service providers develop new services more quickly and cost-effectively, for example. Also, at the highest conceptual level, both IN and TINA propose to address these issues of new service delivery by specifying flexible distributed systems on which the services can be built.

One important difference is that TINA has a much more comprehensive scope. This is especially true in the area of Management. IN, of course, includes an entity called a *Service Management System* (SMS). As normally implemented, however, the SMS mostly takes care of special management issues introduced by the IN itself; for example, the provisioning of customer records and the loading of logic into SCPs. IN as presently conceived does not attempt to address management of the overall network. (That is the scope of a different standards initiative, usually viewed as complementary to IN, known as the TMN.) TINA, by contrast, is intended to provide a distributed architecture suitable for supporting management applications as well as "service" applications within one unified framework.

Another difference is that TINA very explicitly attempts to utilize all the latest and best concepts from recent research and industry practice in the area of distributed computing. So, for example, it is heavily based on concepts of distributed object computing which were perhaps only a gleam in the eye of a few experts when the original development of IN was undertaken. Also, TINA refers to standards for distributed objects (i.e., OMG CORBA), which are intended to serve a wide variety of application needs, telecom applications being just one specialized group. IN, by contrast, implements the necessary distributed computing "transparencies" by relying on the apparatus of Signaling System No. 7, including the remote operations capabilities provided by TCAP and "global title" addressing.

Finally, TINA from the beginning declared itself free of the legacy systems of the telecommunications network. Classical telecom functions such as the establishment of connections are described in TINA in an abstract way, so that in principle they can be realized by a wide variety of physical systems, such as simple Asynchronous Transfer Mode (ATM) switches or cross-connects. Proponents of TINA see a big contrast with IN here. As is clear from the review of IN history in

Chapter 1, IN took the existence of sophisticated telecom switching systems as a given and added a software layer to them to permit interaction with external computers across a distributed control network. TINA hopes to achieve a quantum leap in flexibility by starting with a clean slate.

So if TINA can do everything that IN can do plus more, why not simply utilize TINA from this point forward? While some pioneers will undoubtedly begin gaining experience with the implementation of TINA, the fact is that as of this writing the TINA specifications are still quite new and not entirely complete. Furthermore, in its deliberate effort to isolate itself from legacy systems and the encumbrances of the physical world, TINA leaves much unsaid—such as what specific operating system capabilities, protocol mechanisms, and physical distribution choices are needed to achieve required levels of reliability, real-time performance, and economical operation. So, a lot of work remains to be done by the industry to implement TINA. Meanwhile, other visions of distributed computing (e.g., those based on the Internet/World Wide Web or on proprietary specifications) exist, and TINA will need to clarify for telecom operators, who are very interested in all of these, how its OMG/CORBA-based infrastructure is complementary and critical to their expansion into the world of information networking.

7.3.3   Overview of the TINA architecture

Having perhaps gained some intuitive understanding of the TINA architecture (as we will call the architecture being developed by the TINA Consortium, or TINA-C) from the preceding comparison with IN, we will conclude with a brief architecture overview. It goes without saying that a detailed review of the TINA architecture is beyond the scope of this book. More detailed information is available from, for example, the proceedings of the TINA 95 Conference, held in Melbourne, Australia, in February 1995.

Figure 7.5 shows the overall TINA architecture. Although this figure shows some additional structure, the TINA architecture can be reasonably said to have just three main parts: a Computing Architecture, a Management Architecture, and a Service Architecture. The Management Architecture is further divided into resource components, which model network elements for management purposes, and Resource Management components, which describe management services.

The Computing Architecture, whose chief feature is the specification of a *Distributed Processing Environment* (DPE), is layered on top of various native computing environments (e.g., combinations of computer hardware and operating systems). The Service Architecture pro-

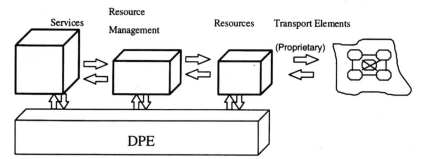

Figure 7.5 TINA layering.

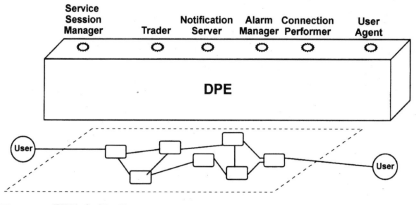

Figure 7.6 TINA-C objective.

vides the means to build telecommunications and information networking services.

As shown in Fig. 7.6, the idea is that, in a physical realization, components (objects) taken from the Service and Management architectures will reside on (possibly) multiple, different physical computing nodes distributed over arbitrary geographical distances. The DPE then provides a "glue" which allows these components to properly interact in a transparent fashion to provide services.

Each of these areas will now be reviewed in a little more detail.

7.3.3.1 Computing Architecture. The Computing Architecture specifies a DPE whose job it is to allow the other architectural components to interact without being aware of such physical realities as geographical distance.

In constructing the Computing Architecture and its DPE, TINA-C relied very heavily on the ODP Reference Model and also on the work of the Object Management Group, including its specification of a *Common Object Request Broker Architecture* (CORBA) and standardized Object Services (Object Management Group, 1994). TINA-C extended the OMG *Interface Definition Language* (IDL) for objects to include, for example, the ability to describe objects with not only "operational" (computing-like) but also "stream" (information flow) interfaces. The result is called the TINA *Object Definition Language* (ODL). However, this is viewed as a strict superset of OMG IDL, and in general the intention of TINA-C is to maintain consistency with OMG.

The DPE physically would be implemented as "DPE kernels" residing on computing nodes in the network, interconnected by a logical network called the *kernel Transport Network* (kTN). The kTN could be realized in a number of ways, such as distinct ATM virtual paths—or possibly even using SS7.

7.3.3.2 Management architecture. The TINA Management Architecture specifies components to support network management, service management, and the management of the DPE itself. The Management Architecture adopts many key principles and concepts from TMN, such as the layered architecture and the idea of managed objects. However, it also integrates ideas from OSI Systems Management and some new capabilities made possible by drawing upon the TINA Computing Architecture and its DPE. Regarding the modeling of resources at the network level, the TINA Network Resource Information Model follows ITU Recommendation G.803 in describing them as composed of subnetworks, connections, termination endpoints, and "trails."

One final important note: going well beyond the traditional definition of "management," the TINA Management Architecture includes fairly detailed specifications for what is called *Connection Management*. The services of Connection Management are intended to actually create end-to-end network connections when called upon, for example, by TINA Service Session Management (which is described in the Service Architecture).

As shown in Fig. 7.7, this is a multilayered architecture which ultimately interfaces with switches through an entity called the Element Management Layer Connection Performer. The TINA Connection Management Architecture appears to provide a lot of flexibility for service and network implementation—though practical engineers who may be reading this are no doubt already thinking about the implications of all those layers for the normally real-time-intensive job of creating network connections.

7.3.3.3 Service architecture. The TINA Service Architecture describes reusable components oriented toward the construction of telecommunications information networking services. These include, for example, Generic Session Endpoints, User Agents, Service Session Managers,

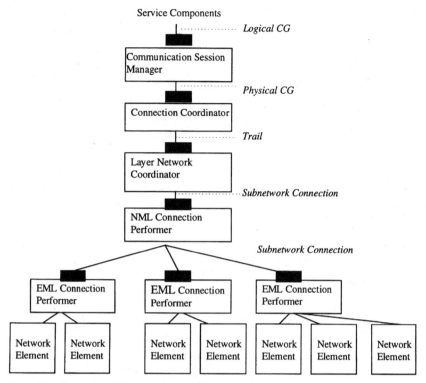

Figure 7.7   Connection Management architecture.

and Communication Session Managers. The TINA Service Architecture model is one in which the Service Session plays a key role. This is the "context" within which parties (end users, subscribers, and service providers) request and negotiate for service capabilities, which in turn will result in the allocation of resources (connections, specialized resources such as conference bridges, etc.). The intention is clearly to support the sort of dynamic, multimedia, multipoint communications sessions which are commonly envisioned to be enabled by the development of broadband ATM networks.

7.4   Regional IN Standards

The material of this section is complementary to the discussion of the standards activities in Chap. 1. Chapter 1 addressed the rise of the regional standards bodies in influencing the ITU-T Recommendations. This section covers the reverse process—adopting the ITU-T Recommendations and turning them into regional standards.

7.4.1 ETSI (Europe)

ETSI NA6 and SPS3 started the work on the European IN standard almost immediately after the initial version of the CS-1 IN Recommendations had been completed.[302] In 1994, the ETSI CS-1 Core INAP, Part 1 (ETSI, 1994) was approved, following the Public Inquiry. The CS-1 Core INAP is a true subset of the ITU-T CS-1 INAP. The subset is DP-generic: it contains none of the DP-specific operations; it prescribes the generic *IntitialDP* operation instead. A more subtle effect of being DP-generic is the replacement of both *AnalyzedInformation* and *SelectRoute* operations with one *Connect* operation.[303] Another CS-1 operation omitted from ETSI Core INAP is *HoldCallinNetwork,* whose semantic role is taken by the *ResetTimer* and *Furnish-ChargingInformation* operations.

ETSI (1994) does not include the SCF-to-SDF protocol, which is to be published in a separate European Telecommunication Standard specification. The rest of the ETSI CS-1 Core INAP specifications (some of which are in the approval process at the moment of this writing) are as follows:

- Part 2: *Protocol Implementation Conformance Statement* (PICS) *Proforma*
- Part 3: *Test Suite Structure and Test Purposes* (TSS&TP)
- Part 4: *Abstract Test Suite* (ATS) and *Partial Protocol Implementation eXtra Information for Testing* (PIXIT) *Proforma*
- Part 5: *Protocol Specification for the SCF-SDF Interface*
- Part 6: *PICS Proforma for the SCF-SDF Interface*

7.4.2 T1S1 (United States)

The ANSI IN Standard project started in T1S1 soon after the first round of work on CS-1 was completed. The goal of the project is to adopt the ITU-T CS-1 Recommendations with the modifications necessary to support the North American implementation. The present schedule (as reported in Spencer and Walsh, 1996) is to submit the standard to the ballot process by the end of 1996 and resolve the comments by the end of 1997. If the project holds to this schedule, the Standard will be ready for ANSI publication in the first quarter of 1997.

---

[302]Note that the work on the ETSI IN Standard has also influenced the CS-1 Refinement, as was reported in the previous chapter.

[303]The *Connect* operation has an effect of these two DP-specific operations performed together.

The major agreements regarding the contents of the Standard are as follows:

- The first ANSI IN Standard will not address the IN Service Plane and Global Functional Plane. As far as the remaining planes are concerned, it has been agreed that ITU-T Recommendations Q.1214, Q.1215, and Q.1225 will form the base of the Standard. The major criterion for adopting (or dropping) any particular aspect of CS-1 is whether or not it can be supported by the existing implementations. One aspect—internetworking—is new, but will be in the Standard based on the CS-1 SCF-to-SDF relationship.

- As far as the ANSI BCSM goes, the original CS-1 BCSM has been enhanced with the addition of new PICs and DPs to the point that it is much closer to the CS-2 BCSM, which is not surprising because the CS-2 BCSM has been built almost entirely on the U.S. contributions.

- T1S1 has decided to support only the DP-specific option of INAP, for which reason DP-generic operations (i.e., *InitialDP* and *Connect*) will not be supported by the Standard. Some subtle changes are also necessary because the ANSI SS No. 7 TCAP is different from the ITU-T TCAP. Additional changes are required to support the interworking between the INAP and ISUP, since the ANSI SS No. 7 ISUP is different from the ITU-T ISUP. As far as the encoding is concerned, T1S1 decided to use the version of ASN.1 specified in ITU-T Recommendation X.680.

7.4.3 CITEL (the Americas)

The Inter-American Telecommunications Commission (better known by its Spanish acronym, CITEL) of the Organization of American States is in charge of the technology coordination across the Americas. One important aspect of its function is making recommendations for adapting existing standards to the needs of the region.[304]

The Working Group For Standards Coordination of the Permanent Consultative Committee I (PCC I) within CITEL created the Subworking group on IN, whose charter is to make the appropriate recommendations regarding IN. The group is considering the ITU-T Recommendations, and is in the process of selecting an appropriate subset INAP. In order to do that, the group has undertaken a study of the present existing IN implementations and implementation plans in the Americas and produced a detailed comparison of the ETSI Core

---

[304]CITEL does *not* write standards itself.

INAP with the Bellcore *Advanced Intelligent Network* (AIN) 0.1 specifications (implemented in the U.S. networks of the Bell Operating Companies). The latest report on the progress of the project can be found in Rinker (1995).

7.4.4 TTC (Japan)

So far, the primary purpose of IN standardization in Japan has been to support roaming for *Personal Handyphone System* (PHS). Currently, TTC has been dealing with two standards concerning IN:

1. JT-Q1218: Internetwork Interface for Intelligent Network
2. JT-Q1218-a: Internetwork Interface for PHS Roaming

The JT-Q1218 standard is mainly based on the revised (i.e., 1995) version of the ITU-T IN CS-1. The specific ITU-T Recommendations used by the standard are Q.1214, Q.1215, Q.1218, and Q.1290. In April 1996, TTC revised the JT-Q1218 to add the SDF-to-SDF interface based on the Draft ITU-T IN CS-2 Recommendations Q.1224 and Q.1228. TTC also plans to develop an application service interface using an internetwork interface based on the ITU-T IN CS-1 SCF-to-SDF interface by the end of 1996.

The JT-Q1218-a standard is based on the emerging ITU-T IN CS-2 Recommendations. To this end, TTC has initiated studies to develop the JT-Q1228 standard, which will be based on ITU-T Recommendations Q.1224, Q.1225, Q.1228, and Q.1290.

In April 1996, TTC revised JT-Q1218-a to add a specification of PHS roaming. IN CS-2 is used to provide the capability to transfer a copy of the PHS service profile from a user's home network when the user first requests location registration after entering a network.

Glossary of Acronyms

AC	Application Context
ACP	Action Control Point
AD	Adjunct
AE	Application Entity
AIN	Advanced Intelligent Network
ANI	Automatic Number Identification
ANSI	American National Standards Institute
AOPL	Application-Oriented Parsing Language
AP	Application Process
API	Application Programming Interface
ASE	Application Service Element
ASN.1	Abstract Syntax Notation One
ATM	Asynchronous Transfer Mode
B-ISDN	Broadband Integrated Services Digital Network
BCM	Basic Call Manager
BCP	Basic Call Process
BCUP	Basic Call-Unrelated Process
BER	Basic Encoding Rules
BNCSM	Basic Non-Call-Associated State Model
BRI	(ISDN) Basic Rate Interface
CASE	Computer-Aided Software Engineering
CC	Calling Card
CCAF	Call Control Agent Function
CCAF+	Call Control Agent Function Plus
CCF	Call Control Function

CCIS	Common Channel Interoffice Signaling
CCITT	International Telephone and Telegraph Consultative Committee
CID	Call Instance Data
CIDFP	Call Instance Data Field Pointer
CITEL	Inter-American Telecommunications Commission
CORBA	Common Object Request Broker Architecture
CPE	Customer Premises Equipment
CPH	Call Party Handling
CRACF	Call-Related Radio Access Control Function
CS	Capability Set
CS-1	Capability Set 1
CS-2	Capability Set 2
CS-3	Capability Set 3
CSM	Call Segment Model
CURACF	Call-Unrelated Radio Access Control Function
CUSF	Call-Unrelated Service Function
CVS	Connection View State(s)
DAF	Distributed Architecture Framework
DFP	IN Distributed Functional Plane
DP	Detection Point
DPE	Distributed Processing Environment
DSDC	Direct Services Dialing Capabilities
DTMF	Dual-Tone Multi-Frequency
ECSA	Exchange Carriers Standards Association
EDP	Event Detection Point
ETSI	European Telecommunications Standards Institute
FCC	Federal Communications Commission
FE	Functional Entity
FEA	Functional Entity Action
FEAM	Functional Entity Access Manager
FIM	Feature Interaction Manager
GFP	(IN) Global Functional Plane
GNS	Green Number Service
GSL	Global Service Logic
HLR	Home Location Register
HLSIB	High-Level SIB
IAF	Intelligent Access Function

IDL	Interface Definition Language
IF	Information Flow
IN-SSM	IN-Switching State Model
IN-WATS	Inward Wide Area Telecommunications Service
INAP	Intelligent Network Application Protocol
INCM	Intelligent Network Conceptual Model
IP	Intelligent Peripheral
ISO	International Organization for Standardization
ISUP	ISDN User Part (Protocol)
ITU	International Telecommunications Union
ITU-T	International Telecommunications Union—Telecommunication Standardization Sector
kTN	kernel Transport Network
LEC	Local Exchange Carrier
LTA	Long Term Architecture
MACF	Multiple Association Control Function
ME-C	Management Entity Control
MTP	Message Transfer Part
N-ISDN	Narrow-band Integrated Services Digital Networks
NA	Network Architecture Technical Committee of ETSI
NAP	Network Access Point
NCP	Network Control Point
NSCX	Network Services Complex
NSP	Network Service (Control) Point
O_BCSM	Originating Basic Call State Model
ODL	(TINA) Object Definition Language
ODP	Open Distributed Processing
OMG	Object Management Group
ONA	Open Network Architecture
OSI	Open Systems Interconnection
OSO	Originating Screening Office
PBX	Public Branch Exchange
PDU	Protocol Data Unit
PE	Physical Entity
PIC	Point in Call
PLMN	Public Land Mobile Network
POI	Point of Initiation
POR	Point of Return

POS	Point of Synchronization
POTS	Plain Old Telephone Service
PRI	(ISDN) Primary Rate Interface
PSPDN	Packet-Switched Public Data Network
PSTN	Public Switched Telephone Network
RBOCs	Regional Bell Operating Companies
RCF	Radio Control Function
ROSE	Remote Operations Service Element
RPC	Remote Procedure Call
SACF	Single Association Control Function
SAO	Single Association Object
SCCP	Signaling Connection Control Part
SCE	Service Creation Environment
SCEF	Service Creation Environment Function
SCEP	Service Creation Environment Point
SCF	Specialized Resource Function
SCME-Control	Service Control Management Entity-Control
SCME-FSM	Service Control Management Entity- Finite State Model
SCP	Service Control Point
SCSM	Service Control Function Call State Model
SCUAF	Service Control User Agent Function
SDF	Service Data Function
SDF-FSM	Service Data Function-Finite State Model
SDL	Specification and Description Language
SDN	Software-Defined Network
SDP	Service Data Point
SIB	Service-Independent Building Block
SMAF	Service Management Agent Function
SMAP	Service Management Agent Point
SMF	Service Management Function
SMP	Service Management Point
SN	Service Node
SPI	Systems Programming Interface
SPS	Signalling Protocols and Switching Technical Committee of ETSI
SRF	Specialized Resource Function
SRME	Specialized Resource Management Entity

SRSM	Specialized Resource Function Call State Model
SS No. 7	Signaling System No. 7
SSCP	Service Switching and Control Point
SSD	Service Support Data
SSF	Service Switching Function
SSF-FSM	Service Switching Function-Finite State Model
SSME	Service Switching Function Management Entity
SSME-Control	Service Switching Function Management Entity-Control
SSP	Service Switching Point
STM	Synchronous Transfer Mode
T_BCSM	Terminating Basic Call State Model
T1S1	T1S1 Committee
TASM	Terminal Access State Model
TCAP	Transaction Capabilities Application Part
TDP	Trigger Detection Point
TDP	Trigger Detection Point (a statically armed DP)
TINA	Telecommunications Information Network Architecture
TINA-C	Telecommunications Information Network Architecture Consortium
TMN	Telecommunications Management Network
TTC	Telecommunications Technology Committee
UFM	Unified Functional Methodology
UPT	Universal Personal Telecommunications
US ITAC	United States International Telecommunications Advisory Committee
VLR	Visitor Location Register
VPN	Virtual Private Network
WIN	Wireless Intelligent Network

Bibliography

Ai, Bo, Bai Wang, and Jun-Liang Chen. 1995: "CIN01—The IN Implementation in China," *Record of the IEEE IN '95 Workshop,* May 9–11, Ottawa, Canada.

Akihara, Masaya, K. Shimizu, and S. Ito. 1995: "An Implementation and Evaluation of Service Creation Environment based on ITU-T IN CS-1," *Record of the IEEE IN '95 Workshop,* May 9–11, Ottawa, Canada.

Andrews, F. T., Jr. and K. E. Martersteck. 1982: Prologue to Special Issue on Stored Program Control Network, *Bell System Technical Journal* **61**(7, Part 3): 1575–1577.

Ambrosch, Wolf-Dietrich, Anthony Maher, and Barry Sasscer (Eds.). 1989: *The Intelligent Network,* Berlin and Heidelberg: Springer-Verlag.

Asmuth, R. L., and G. W. Gawrys. 1981: *Direct Services Dialing Capabilities Specification Document FSD81-03T,* Internal AT&T Document.

AT&T. 1985: *AT&T Today Began Offering Three Major New Services,* Press Release, November 4.

Baer, K., et al. 1988: "Evolution of the Intelligent Network," *Proceedings of the National Communications Forum* **42**: 444–451.

Barr, William J., Trevor Boyd, and Yuji Inoue. 1993: "The TINA Initiative," *IEEE Communications Magazine* **31**(3): 70–76.

Bassinger, R. G., M. Berger, E. M. Prell, V. R. Ransom, and J. R. Williams. 1982: "Calling Card Service—Overall Description and Operational Characteristics," *Bell System Technical Journal* **61**(7, Part 3): 1655–1673.

Bennett, Ronnie L., Jerry C. Chen, and Thomas B. Morawski. 1991: "Intelligent Network OAM&P Capabilities and Evolutions for Network Elements," *AT&T Technical Journal* **70**(3): 85–98.

Berman, R. K., and J. H. Brewster. 1992: "Perspectives on the AIN Architecture," *IEEE Communications Magazine* **30**(2): 27–32.

Cellular Telecommunications Industry Association. 1996: "Standards Requirements Document for the Wireless Intelligent Network: MSC and Associated Interfaces," Draft, April.

Chang, Ty. 1994: "Discussing the Weaknesses of the Standard Service-Independent Building Blocks," *Proceedings of the Third International Conference on Intelligence in Networks (ICIN94),* October 11–13, Bordeaux, France.

Clarisse, O. B. E. A., Kidwell, W. F. Opdyke, R. E. Pitt, and B. A Westergren. 1994: "Service Creation Using Application Building Blocks," October 1994, *Proceedings of the Third International Conference on Intelligence in Networks (ICIN94),* October 11–13, Bordeaux, France.

Dacloush, Elias Joseph, I. Faynberg, L. R. Gabuzda, and W. Huen. 1995: "On Intelligent Network Application Protocol (INAP) and Relevant Implementation Issues," *Proceedings of the International Symposium on Intelligence in Broadband Networks (ISIB '95),* April 10–15, Beijing, People's Republic of China.

Daryani, Pramila, I. Faynberg, S. J. Griesmer, M. P. Kaplan, and A. L. Waxman. 1992: "Object-Oriented Modelling of the Intelligent Network and Its Application to Universal Personal Telecommunications Service," *Proceedings of the International Council for Computer Communications Intelligent Networks Conference,* May 4–6, Tampa, Fla.

Elmgren, Maria, and H. Majeed. 1994: "Telecom Australia's and Ericsson Australia's Experience Implementing Line-Based Services with the IN SIB Concept," *Record of the IEEE IN '94 Workshop,* May 24–26, Heidelberg, Germany.

ETSI. 1994: Intelligent Network Capability Set 1 (CS1) Core Intelligent Network Application Protocol (INAP) Part 1: Protocol specification, European Telecommunication Standard ETS **300** 374–1, September, Sophia Antipolis, France: European Telecommunications Standards Institute.

Faynberg, Igor, Lawrence R. Gabuzda, M. P. Kaplan, M. V. Kolipakam, W. J. Rowe, and A. L. Waxman. 1992: "The Support of Network Interworking and Distributed Context Switching in the IN Service Data Function Model," *Proceedings of the Second International Conference on Intelligence in Networks,* March 3–5, Bordeaux, France.

Faynberg, Igor, Lawrence R. Gabuzda, and Doris S. Lebovits. 1993: "On Intelligent Network Architectural Concepts and their Possible Evolution and Application to TINA," *Proceedings of the Fourth Telecommunications Information Networking Architecture Workshop,* September 27–30, L'Aquila, Italy.

Faynberg, Igor. 1995: "On Internetwork Access in IN Standards (Current status and possible options for the future)," Presented at AIN ComForum, February 7–8, Phoenix, Ariz.

Faynberg, Igor. 1995: Marked-up Version of Draft Recommendation Q.1215. *ITU-Telecommunication Standardization Sector,* Study Group 11, Working Party 4, Plenary Document TD PL/11-11, May 5, Geneva.

Grinberg, Arkady. 1995: *Computer / Telecom Integration: the SCAI Solution.* New York: McGraw-Hill.

Gulzar, Zaman, and Frank Salm. 1995: "Developing AIN Products Using AIN 2OOL-Kit™," *Record of the IEEE IN '95 Workshop,* May 9–11, Ottawa, Canada.

Haas, C. W., D. C. Salerno, and D. Sheinbein. 1982: "800 Service Using SPC Network Capability-Network Implementation and Administrative Functions," *Bell System Technical Journal* **61**(7, Part 3): 1745–1757.

Hass, R. J., and R. B. Robrock. 1986: "Introducing the Intelligent Network," *Bellcore EXCHANGE* July/August: 3–7.

Hetz, Harry A., and Timothy G. Rinker. 1995: "International and Domestic IN Standardization Process and Status," Presented at *AIN ComForum,* February 7–8, Phoenix, Ariz.

Hilton, Jay R. 1996: "Q.1222: IN CS-2 Service Plane—Editor's Draft Text," *ITU-Telecommunication Standardization Sector,* Study Group 11, Working Party 4, TD 4/11—33, January 29, Miyazaki, Japan.

Holzmann, Gerard J. 1991: *Design and Validation of Computer Protocols,* Englewood Cliffs, N.J.: Prentice-Hall.

Horing, S., J. Z. Menard, R. E. Staehler, and B. J. Yokelson. 1982: "Stored Program Control Network—Overview," *Bell System Technical Journal* **61**(7, Part 3):1579–1588.

Hu, Yun-Chao, 1995: Marked-up Version [of] Draft Q.1213 Recommendation, *ITU-Telecommunication Standardization Sector,* Study Group 11, Working Party 4, Plenary Document TD PL/11-17, May 5, Geneva.

Hu, Yun-Chao, and A. Herian. February 1996: "Draft Q.1223 Recommendation," *ITU-Telecommunication Standardization Sector,* Study Group 11, Working Party 4, TD 4/11-44, February 9, Miyazaki, Japan (the original version augmented with the editoral changes agreed on at the meeting).

Isidoro, Allesandro L. 1992: "Beyond the Two-Party Call: A Methodology Based on Call Configurations," T1S1.1/92-256R1, Contribution to Working Group T1S1.1.

ISO 10746-2 RM-ODP Part 2. June 1993: Descriptive Model, Yokahama.

ITU-T Z.100. 1994: Recommendation Z.100, "Specification and Description Language (JDL)," International Telecommunications Union Standardization Section, Geneva.

Kettler, Herbert W., Gautham Natarajan, Ecaterina W. Scher, Philip Y. Shih, and Peggy Marr Wainscott. 1991: "AT&T's Global Intelligent Network Architecture," *AT&T Technical Journal* **71**(5): 30–35.

Ku, Bernard S. 1994: "Service Creation Environment in Action," *Proceedings of the Third International Conference on Intelligence in Networks (ICIN94)*, October 11–13, Bordeaux, France.

Kung, R., and E. Paul. 1995: "The Intelligent Network," *Commutation & Transmission* **2**: 5–12.

Lawser, J. J., R. E. LeCronier, and R. L. Simms. 1982: "Stored Program Control Network—Generic Network Plan," *Bell System Technical Journal* **61**(7, Part 3): 1589–1598.

Lawser, John J., and Daniel Sheinbein. 1979: "Realizing the Potential of the Stored Program-Controlled Network," *Bell Laboratories Record* **57**(3): 85–89.

Lu, Hui-Lan. February, 1996: "Draft ITU-T Recommendation Q.1221," *ITU-Telecommunication Standardization Sector,* Study Group 11, Working Party 4, TD 4/11–18, January 29, Miyazaki, Japan.

Mano, M. Morris. 1979: *Digital Logic and Computer Design,* Englewood Cliffs, N.J.: Prentice-Hall.

Marks, Joel M., Cameron L. Wolff, and Katherine A. Koenig. 1995: "Service Programmability in the Intelligent Network," *Record of the IEEE IN '95 Workshop,* May 9–11, Ottawa, Canada.

Mearns, Allison B., David J. Miller, and Cyrenus M. Rubald. 1982: "Calling Card: Don't Tell It—Dial It," *Bell Laboratories Record,* May/June: 117–119.

Members of Technical Staff and the Technical Publications Department, AT&T Bell Laboratories. 1986: *Engineering and Operations in the Bell System,* 2nd Ed., Murray Hill, N.J.: AT&T Bell Laboratories.

Mitra, Nilotpal, and S. Usiskin. 1995: "Interrelationship of the SS7 Protocol Architecture and the OSI Reference Model and Protocols," *The Froehlich/Kent Encyclopedia of Telecommunications,* **9**:491–519. New York: Marcel Dekker, Inc.

Mitra, Nilotpal. 1995: "An Introduction to the ASN.1 MACRO Replacement Notation," *AT&T Technical Journal* **73**(3): 66–79.

Mitra, Nilotpal. 1995: "Efficient Encoding Rules for ASN.1-Based Protocols," *AT&T Technical Journal* **73**(3): 80–93.

Morgan, Michael J., Michael J. Cosky, Thomas M. Gruenenfelder, T. Curtis Holmes, Jr.,and Gerald A. Raack. 1991: "Service Creation Technologies for the Intelligent Network," *AT&T Technical Journal* **70**(3): 58–71.

Olsen, Anders, O. Faergemand, B. Møller-Pedersen, R. Reed, and J. R. W. Smith. 1994: *Systems Engineering Using SDL-92,* Amsterdam, the Netherlands: Elsevier Science B.V.

Object Management Group, January 1994: "Common Object Services Specifications," OMG Document, Vol. 1.

Omiya, Tomoki, and Takashi Suzuki. 1994: "Architecture and Interfaces for the Introduction of Advanced IN," *Proceedings of the Third International Conference on Intelligence in Networks (ICIN94),* October 11–13, Bordeaux, France.

Peterson, James L., and Abraham Silberschatz. 1985: *Operating System Concepts,* 2d Ed., Reading, Mass.: Addison-Wesley.

Petruk, Martin H. 1994: "Experiences in Applying the ITU Capability Set 1 Intelligent Network Conceptual Model (INCM) to Service Creation," *Proceedings of the Third International Conference on Intelligence in Networks (ICIN94),* October 11–13, Bordeaux, France.

Reiner, Mascall, Anthony J. Drignath, and Eckhard Speller. 1994: "Alcatel's Intelligent Network Product Line," *Record of the IEEE IN '94 Workshop,* May 24–26, Heidelberg, Germany.

Rinker, Timothy G. 1995: "Chair's Report on the Meeting of the Working Group for Standards Coordination," Permanent Consultative Committee I, Inter-American Telecommunications Commission, Organization of American States, WGSC-D0007-E. September 1, Washington, D.C.

Ritchie, A. E., and J. Z. Menard. 1978: "Common Channel Interoffice Signaling—an Overview," *Bell System Technical Journal* **57**(2): 221–224.

Ritchie, A. E., and L. S. Tuomenoksa. 1977: "No. 4ESS—System Objectives and Organization," *Bell System Technical Journal* **56**(7): 1017–1027.

Ruggles, Clive L. N., Ed. 1990: Formal Methods in Standards: A Report from the BCS Working Group, Norwich, Great Britain: Springer-Verlag.

Russo, E. G., M. Tawfik Abdel-Moneim, Linda L. Sand, Laura J. Shaw, Patricia D. Taska, and Ronald J. Wojcik. 1991: "Intelligent Network Platforms in the U.S.," *AT&T Technical Journal* **70**(3): 26–43.

Sable, E. G., and Herbert W. Kettler. 1991: "Intelligent Network Directions," *AT&T Technical Journal* **70**(3): 2–10.

Sheinbein, D., and R. P. Weber. 1982: "800 Service Using SPC Network Capability," *Bell System Technical Journal* **61**(7, Part 3): 1737–1757.

Slutsman, Lev, Hui-Lan Lu, Mark P. Kaplan, and Igor Faynberg. 1994: "The Application-Oriented Parsing Language (AOPL) as a Way to Achieve Platform-Independent Service Creation Environment," *Proceedings of the Third International Conference on Intelligence in Networks (ICIN94),* October 11–13, Bordeaux, France.

Spencer, Cliffton. 1995: "Service Creation Administrative Layer," *Record of the IEEE IN '95 Workshop,* May 9–11, Ottawa, Canada.

Spencer, Cliffton, and Thomas D. Walsh. 1996: "The Emerging ANSI IN Standard based upon ITU-T Recommendations," *Record of the IEEE IN '96 Workshop,* April 21–24, Melbourne, Australia.

Squires, Dennis N. 1995: "Implementing IN Services through SPACE®," *Record of the IEEE IN '95 Workshop,* May 9–11, Ottawa, Canada.

Stallings, William. 1992: *ISDN and Broadband ISDN,* Englewood Cliffs, N.J.: Macmillian Publishing Company.

Suzuki, Shigehiko. 1993: "An Overview of the Progress in the Intelligent Network," *NTT Review* **5**(5): 32–37.

Sykes, Jack S., and Eldred J. Visser. 1992: "Globalization of Intelligent Network Services," *AT&T Technical Journal* **71**(5): 6–12.

Syrett, Mark, David Skov, and Anders Kristensen. 1995: "HP in IN," *Record of the IEEE IN '95 Workshop,* May 9–11, Ottawa, Canada.

Tanenbaum, Andrew S. 1989: *Computer Networks.* 2d Ed., Englewood Cliffs, N.J.: Prentice-Hall.

Tanenbaum, Andrew S. 1992: *Modern Operating Systems,* Englewood Cliffs, N.J.: Prentice-Hall.

Thieffry, Jean-Bernard. 1995: Revised ITU-T Recommendation Q.1214-Distributed Functional Plane for Intelligent Network CS-1 (Parts I and II), *ITU-Telecommunication Standardization Sector,* Study Group 11, Working Party 4, Plenary Documents TD PL/11-11 and TD PL/11-12, May 5, Geneva.

Turner, Douglas. May 1995: Q.1218-Final Edited Copy (Parts 1-3). *ITU-Telecommunication Standardization Sector,* Study Group 11, Working Party 4, Plenary Documents TD PL/11-14, TD PL/11-15, and TD PL/11-16, May 5, Geneva.

Ullrich, Rita. 1984: "Customers Tailor Call Routes with Advanced 800 Service," *Bell Labs News,* December 10, p. 1.

Visser, John. 1991: "Framework for Intelligent Network Concepts and Standards," *Proceedings of the International Conference on Communications ICC 91,* Vol. 3: 1277–1282, June 23–27, Denver, Colo.

Workman, Alexandra M., Murthy V. Kolipakam, Janis B. Sharpless, Vilma Stoss, and Hans van der Veer. 1991: "International Applications of AT&T's Intelligent Network Platforms," *AT&T Technical Journal* **70**(3): 44–57.

Wyatt, George Y., Alvin Barshefsky, Robert V. Epley, Marc P. Kaplan, and Krish P. Krishnan. 1991: "The Evolution of Global Intelligent Network Architecture," *AT&T Technical Journal* **70**(3): 11–25.

Zeuch, Wayne R. 1996: Revised Recommendation Q.65: "The Unified Functional Methodology for the Characterization of Services and Network Capabilities," *ITU-T Question 1/11,* TD PL-11/41 R1, February, Miyazaki, Japan.

Index

ABOUT THE AUTHORS

IGOR FAYNBERG is a Member of the Technical Staff in the Global Strategic Standardization Department of Bell Laboratories where he leads the work on IN Standards and their applications to the needs of various businesses. In the past six years Dr. Faynberg has been actively contributing to the American National Standards Institute (ANSI) T1S1 committee and the International Telecommunications Union where he is presently the General Editor of ITU-T (formerly CCITT) IN Recommendations Series. Dr. Faynberg has been assigned by the ANSI committee T1S1 to represent it as the Liaison Rapporteur at the meetings of European Telecommunications Standards Institute (ETSI). Dr. Faynberg holds Mathematics Diploma (1975) from Kharkov University, Ukraine, and M.S. (1984) and Ph.D. (1989), both in Computer and Information Science, from the University of Pennsylvania.

LAWRENCE R. GABUZDA is currently a Technical Manager in the Global Strategic Standardization Department of Bell Laboratories, where his responsibilities include the application of Intelligent Network control principles and distributed computing technology to the development of advanced network control architectures supporting emerging personal communications and multimedia services. Prior to assuming his current position, Mr. Gabuzda had a career of over ten years with the Switching Systems Engineering Division of Bell Laboratories, where he made various contributions to the evolution of the 4ESS™ and 5ESS® switching systems, Intelligent Network products, and AT&T Intelligent-Network-based services. Mr. Gabuzda holds a B.S.E.E. from Rensselaer Polytechnic Institute, M.S.E.E. from Carnegie Mellon University, and M.S. in Systems Engineering from the University of Pennsylvania.

MARC P. KAPLAN is the Technical Manager responsible for the planning and developing service support systems for some of AT&T's newest business communications services. During his career at Bell Labs Mr. Kaplan has supervised the InBound (800/900) Service Creation Prototyping Group at AT&T Bell Laboratories; coordinated direction and developed corporate consensus for external participation in network standards; participated in ITU-T (formerly CCITT) and T1S1 IN standards, serving as an editor for ITU-T Recommendations Q.1201 and Q.1211; led an AT&T technology assessment of the role of Common Channel Signaling System No. 7(CCS7) in premises products and the consequences for network architecture and network services; participated in cooperative-planning activities between AT&T Bell Laboratories, Bellcore, and several Regional Bell Operating Companies on IN2; led the technical specification teams for AT&T MEGACOM® 800 and 800 READYLINE®; managed the project of an 800 Service upgrade; and provided technical consulting and participated in initial network design and equipment requirements of British Telecom's DDSN for FreePhone service. Mr. Kaplan holds a B.S. Eng. in Systems Engineering from the University of Illinois (Chicago) and an M.S.C.E. from Northwestern University.

NITIN J. SHAH is Technology Director of the Wireless Core Technology Department in Lucent Technologies Network Wireless systems. His group is part of the network infrastructure manufacturing division of AT&T responsible for cellular, personal communications, and other wireless communications products. Dr. Shah has responsibility for technology and architecture for wireless networks. His organization has projects on technology planning, wireless network architecture, radio multiple access technologies, and digital compression technologies for speech and visual communication. Dr. Shah received his BA (1979), M.A. (1982), and Ph.D. (1983) in Microelectronic Engineering, all from the University of Cambridge, in Cambridge, England.